Methods of
Experimental Physics

VOLUME 9
PLASMA PHYSICS
PART B

METHODS OF EXPERIMENTAL PHYSICS

L. Marton, *Editor-in-Chief*

Claire Marton, *Assistant Editor*

1. Classical Methods
   Edited by Immanuel Estermann

2. Electronic Methods
   Edited by E. Bleuler and R. O. Haxby

3. Molecular Physics
   Edited by Dudley Williams

4. Atomic and Electron Physics—Part A: Atomic Sources and Detectors; Part B: Free Atoms
   Edited by Vernon W. Hughes and Howard L. Schultz

5. Nuclear Physics (*in two parts*)
   Edited by Luke C. L. Yuan and Chien-Shiung Wu

6. Solid State Physics (*in two parts*)
   Edited by K. Lark-Horovitz and Vivian A. Johnson

7. Atomic and Electron Physics—Atomic Interactions (*in two parts*)
   Edited by Benjamin Bederson and Wade L. Fite

8. Problems and Solutions for Students
   Edited by L. Marton and W. F. Hornyak

9. Plasma Physics (*in two parts*)
   Edited by Hans R. Griem and Ralph H. Lovberg

Planned volumes:—
   Optical Methods
   Edited by R. M. Scott and B. F. Thompson

   Acquisition and Analysis of Data
   Edited by L. J. Kieffer

Volume 9

Plasma Physics

PART B

Part B *Edited by*

RALPH H. LOVBERG

*Department of Physics, Institute of Geophysics and Planetary Physics
University of California at San Diego
La Jolla, California*

HANS R. GRIEM

*Department of Physics and Astronomy
University of Maryland
College Park, Maryland*

1971

ACADEMIC PRESS • New York and London

COPYRIGHT © 1971, BY ACADEMIC PRESS, INC.
ALL RIGHTS RESERVED
NO PART OF THIS BOOK MAY BE REPRODUCED IN ANY FORM,
BY PHOTOSTAT, MICROFILM, RETRIEVAL SYSTEM, OR ANY
OTHER MEANS, WITHOUT WRITTEN PERMISSION FROM
THE PUBLISHERS.

ACADEMIC PRESS, INC.
111 Fifth Avenue, New York, New York 10003

United Kingdom Edition published by
ACADEMIC PRESS, INC. (LONDON) LTD.
Berkeley Square House, London W1X 6BA

LIBRARY OF CONGRESS CATALOG CARD NUMBER: 68-56970

PRINTED IN THE UNITED STATES OF AMERICA

CONTENTS OF VOLUME 9, PART B

CONTRIBUTORS TO VOLUME 9, PART B	vii
FOREWORD	ix
PREFACE TO VOLUME 9	xi
CONTENTS OF VOLUME 9, PART A	xiii
CONTRIBUTORS TO VOLUME 9, PART A	xvii
LIST OF SYMBOLS	xix

11. Optical Refractivity of Plasmas

by F. C. JAHODA AND G. A. SAWYER

11.1. Introduction	1
11.2. Theory	3
11.3. Standard Experimental Methods	13
11.4. Experimental Methods Requiring Laser Sources . . .	31

12. Deep Space Plasma Measurements

by VYTENIS M. VASYLIUNAS

12.1. Introduction	49
12.2. Instrumentation	52
12.3. Methods of Analysis	77

13. Whistlers: Diagnostic Tools in Space Plasma

by NEIL M. BRICE AND ROBERT L. SMITH

13.1. Introduction	89
13.2. Experimental Method	90
13.3. Ground-Based Whistler Observations	92
13.4. Satellite Observations	111
Appendix	135

14. Radio Wave Scattering from the Ionosphere
by D. T. FARLEY

14.1. Introduction	139
14.2. Scattering from a Diffuse Medium	142
14.3. Incoherent Scattering	158
14.4. Scattering from the Equatorial Electrojet	176

15. Dense Plasma Focus
by J. W. MATHER

15.1. Introduction	187
15.2. The Apparatus	190
15.3. Plasma Focus Development	194
15.4. Plasma Diagnostic Measurements	215
15.5. Summary	248

16. Plasma Problems in Electrical Propulsion
by RALPH H. LOVBERG

16.1. Introduction	251
16.2. Electromagnetic Propulsion as a Problem in Magnetohydrodynamics (MHD)	255
16.3. Electric Propulsion as a Plasma Physics Problem: The Magnetoplasmadynamic Arc	269

AUTHOR INDEX	291
SUBJECT INDEX	297

CONTRIBUTORS TO VOLUME 9, PART B

Numbers in parentheses indicate the pages on which the authors' contributions begin.

NEIL M. BRICE, *School of Electrical Engineering, Cornell University, Ithaca, New York* (89)

D. T. FARLEY, *Center for Radiophysics and Space Research, Cornell University, Ithaca, New York* (139)

F. C. JAHODA, *Los Alamos Scientific Laboratory, University of California, Los Alamos, New Mexico* (1)

RALPH H. LOVBERG, *Department of Physics, Institute of Geophysics and Planetary Physics; University of California at San Diego, La Jolla, California* (251)

J. W. MATHER, *Los Alamos Scientific Laboratory, University of California, Los Alamos, New Mexico* (187)

G. A. SAWYER, *Los Alamos Scientific Laboratory, University of California, Los Alamos, New Mexico* (1)

ROBERT L. SMITH (89), *Radioscience Laboratory, Stanford Electronics Laboratories, Stanford, California* (89)

VYTENIS M. VASYLIUNAS, *Department of Physics and Center for Space Research, Massachusetts Institute of Technology, Cambridge, Massachusetts* (49)

FOREWORD

Volume 9 fulfills the plan stated in the Foreword to Volume 7 to cover various aspects of plasma physics. As in the past, the wealth of material requires two books for its presentation.

Currently planned are volumes dealing with optics and the use and handling of data.

All this effort would not be possible without the devotion of the volume editors and of the numerous contributors, as well as the full cooperation of the publishers. Ample demonstration of this spirit is given in the volumes on hand and it is my pleasant task to thank Professors Griem and Lovberg and all their contributing authors for what is hoped to be a most useful compendium of methods used by the experimental plasma physicist.

L. MARTON

PREFACE TO VOLUME 9

Plasma physics methods find applications in a wide range of disciplines, from astrophysics to electrical power generation. We therefore considered it desirable to combine in Part A (edited by Hans R. Griem) a discussion of the most widely used experimental methods in our field with examples for their applications to laboratory plasma (mostly stability) problems. Part B (edited by Ralph H. Lovberg) mainly summarizes how plasma methods help in investigations of our environment and in technology. There are also chapters dealing with some recent developments in plasma generation, radiation, and optical measurements.

The reader will probably be surprised by the large share of theory in many of the parts. This seems to be typical of plasma physics, and is perhaps unavoidable when dealing with many-body systems (albeit here exclusively classical ones). Experimental *techniques* have already been covered in great detail in other multiauthor books ("Plasma Diagnostic Techniques," S. L. Leonard and R. H. Huddlestone, eds., Academic Press, New York, 1965; "Plasma Diagnostics," W. Lochte-Holtgreven, ed., North-Holland Publ., Amsterdam, 1968) and in some monographs ("Plasma Diagnostics with Microwaves," M. A. Heald and C. B. Wharton, Wiley, New York, 1965; "Plasma Spectroscopy," H. R. Griem, McGraw-Hill, New York, 1964; "Radiation Processes in Plasmas," G. Bekefi, Wiley, New York, 1966). We therefore emphasize general *methods* used in experimental investigations of plasma problems.

Another compromise had to be found in regard to the ideal of a unified notation. Had the editors insisted on an entirely uniform notation throughout the volume, many of the parts would have deviated substantially from general usage in the relevant literature. As one of the foremost functions of this series is to introduce the reader to the more specialized literature, this would have been too high a price to pay. We therefore hope that the appended list of symbols will help the reader to find his way through the volume in spite of multiple meanings of some symbols, and different symbols for the same physical quantity in separate parts.

The choice of systems of units also becomes difficult in an enterprise such as this, since plasma technology, more than most other fields, involves a large overlapping of pure science and engineering. An inevitable result is that the plasma physicist is forced into some kind of compromise between the cgs and mks systems of units (especially in the associated electromagnetic problems), and the point at which this compromise is made will vary from one worker to another.

We thank all contributors for their shares in this task, and Dr. L. Marton, the editor of these Methods volumes, and Academic Press for their steady encouragement.

HANS R. GRIEM

RALPH H. LOVBERG

CONTENTS OF VOLUME 9, PART A
Edited by HANS R. GRIEM

1. **Plasma Waves and Echoes**

    by K. W. GENTLE

    1.1. Plasma Waves
    1.2. Observation of Plasma Waves
    1.3. Plasma Wave Echoes
    1.4. Observations of Echoes

2. **Microwave Scattering from Plasmas**

    by T. C. MARSHALL

    2.1. Introduction
    2.2. The Electromagnetic Problem
    2.3. The Fluctuation Problem
    2.4. The Experimental Problem

3. **Plasma Diagnostics by Light Scattering**

    by A. W. DeSILVA and G. C. GOLDENBAUM

    3.1. Introduction
    3.2. Theory of Light Scattering
    3.3. Major Design Parameters of a Scattering Experiment
    3.4. Optical Systems
    3.5. Interpretation of Data
    3.6. Appendix

4. **Atomic Processes**

    by R. C. ELTON

    4.1. Introduction
    4.2. Experimental Methods of Plasma Analysis
    4.3. Specific Atomic Processes in Plasmas
    4.4. Useful Supplementary Material

5. Plasma Heating by Strong Shock Waves

by R. A. Gross and B. Miller

5.1. Introduction
5.2. Experimental Components of Electromagnetic Shock Tubes
5.3. System Performance and Experimental Results

6. Collisionless Shock Waves in Laboratory Plasmas

by E. Hintz

6.1. Introduction
6.2. Theoretical Models for the Formation and Structure of Shock Waves
6.3. Experimental Methods
6.4. Experimental Results

7. High-Frequency Instabilities

by W. A. Perkins

7.1. Introduction
7.2. Driving Mechanisms for High-Frequency Instabilities
7.3. Types of Distributions That Are Unstable
7.4. Linear Theory and Dispersion Relation
7.5. Absolute and Convective Instabilities
7.6. Longitudinal (Electrostatic) Instabilities
7.7. Transverse (Electromagnetic) Instabilities
7.8. Computer Simulation Experiments

8. Low-Frequency Instabilities

by S. Yoshikawa

8.1. Classification of Low-Frequency Instabilities
8.2. Growth Rates of Low-Frequency Instabilities
8.3. Relation to Experiment
8.4. Experiments
8.5. Experimental Methods To Study Low-Frequency Oscillations
8.6. Conclusions

9. Collisional Drift Instabilities

by H. W. HENDEL AND T. K. CHU

9.1. Introduction
9.2. Theory
9.3. Experiment
9.4. Instability Identification
9.5. Mode Stabilizations
9.6. Critical Fluctuations
9.7. Amplitudes in Stable and Unstable Regimes
9.8. Enhanced Radial Plasma Transport Caused by Collisional Drift Waves
9.9. Concluding Remarks

10. Instabilities of High-Beta Plasmas

by H. A. BODIN

10.1. Introduction
10.2. Examples of High-Beta Systems—Methods of Plasma Production
10.3. Theory
10.4. Diagnostic Measurements
10.5. Experimental Investigation of High-Beta Instabilities
10.6. Dynamic and Feedback Stabilization
10.7. Future Trends

AUTHOR INDEX—SUBJECT INDEX

CONTRIBUTORS TO VOLUME 9, PART A

H. A. B. BODIN, *Culham Laboratory, Abingdon, Berkshire, England*

T. K. CHU, *Plasma Physics Laboratory, Princeton University, Princeton, New Jersey*

A. W. DESILVA, *Department of Physics and Astronomy, University of Maryland, College Park, Maryland*

R. C. ELTON, *Plasma Physics Division, U.S. Naval Research Laboratory, Washington, D.C.*

K. W. GENTLE, *Department of Physics, The University of Texas, Austin, Texas*

G. C. GOLDENBAUM, *Plasma Physics Division, U.S. Naval Research Laboratory, Washington, D.C.*

R. A. GROSS, *Plasma Laboratory, Columbia University, New York, New York*

H. W. HENDEL, *Plasma Physics Laboratory, Princeton University, Princeton, New Jersey*

E. HINTZ, *Institut für Plasmaphysik der Kernforschungsanlage Jülich, Germany*

T. C. MARSHALL, *Department of Electrical Engineering, Columbia University, New York*

B. MILLER, *Department of Nuclear Engineering, Ohio State University, Columbus, Ohio*

W. A. PERKINS, *Radiation Laboratory, University of California, Berkeley, California*

S. YOSHIKAWA, *Plasma Physics Laboratory, Princeton University, Princeton, New Jersey*

LIST OF MOST FREQUENTLY USED SYMBOLS

B	Magnetic field strength	v_d (v_D)	Drift velocity
c_s	Sound velocity	v_e (v_i)	Electron (ion) thermal velocity
D	Diffusion coefficient, Fokker–Planck operator	v_g	Group velocity
E	Electric field strength, excitation energy, "Etendue"	v_p	Phase velocity
		α	Scattering parameter
		β	Ratio of particle and total (magnetic plus particle) pressures
f	Distribution function, oscillator strength		
F	(Ion) distribution function, flux	γ	Growth rate, specific heat ratio, molecular polarizability, relativistic meter
g (G)	Acceleration, statistical weight		
G	Gravitational potential	ε	Dielectric constant or function
h	Enthalpy, Planck's constant	ε_ω	Emission coefficient
I	Current, intensity	η	Compression ratio, quantum efficiency
j	Current density		
k	Wave number, Boltzmann constant	κ	Inverse of plasma scale length
K	Boltzmann constant	$\kappa_{s\omega}$	Absorption coefficient
k_ω	Absorption coefficient	λ	Wavelength, mean free path
m	Mode number, electron mass	λ_D	Debye length
		μ	Permeability, viscosity, refractive index, shear
m_e (m_i)	Electron (ion) mass	ν	Collision frequency
M	Ion mass, Mach number	ξ	Amplitude of instability
n	Particle density, principal quantum number	ρ	Mass density
		σ	Conductivity, cross section
N	Particle density, whole number	ϕ	Potential
		χ	Ionization energy
r	Radius, reflectivity	ω	(Angular) frequency
r_e (r_i)	Electron (ion) gyro radius	ω_{ce} (ω_{ci})	Electron (ion) cyclotron frequency
R (R_c)	Radius of curvature, distance	ω_{pe} (ω_{pi})	Electron (ion) plasma frequency
S	Spectral function, species		
T_e (T_i)	Electron (ion) temperature	ω_i	Statistical weight
u	Fluid velocity, shock velocity, energy density	ω_i^*	(Ion) drift frequency
		Ω	Solid angle, angular velocity, collision strength
v	Particle velocity		
V_A	Alfvén velocity		

Methods of Experimental Physics

VOLUME 9
PLASMA PHYSICS
PART B

11. OPTICAL REFRACTIVITY OF PLASMAS*

11.1. Introduction

In this part we shall be concerned with the propagation through plasmas of electromagnetic waves whose frequencies (optical and near-optical spectral band) are very much larger than any characteristic plasma frequency. In this limit the interaction between wave and plasma is weak and is manifested primarily in a slightly altered propagation velocity of the wave. The ratio of the phase velocity in vacuum to that in the plasma is called the refractive index. In a plasma, the refractive index is primarily a function of electron density, which is the main plasma parameter determined by the various measurements of the phase changes induced in a wavefront as a result of its traversal of a plasma. We limit ourselves to these refractive effects and do not consider the important diagnostic techniques associated either with the very much weaker wave scattering phenomena (Volume 9A, Part 3), or with the emission of electromagnetic radiation by the plasma itself (Volume 9A, Part 4).

Probing with optical radiation is a desirable diagnostic method because the interaction is weak and the plasma conditions will not be perturbed by the measurement itself. By the same token, however, the sensitivity is limited and the usefulness of optical refractivity measurements is generally restricted to plasma densities greater than $\sim 10^{14}/cm^3$. An increasing variety of laboratory produced plasmas, many of them products of research on controlled thermonuclear fusion, fall into this category. Another advantage of optical probing is that the small size of optical wavelengths permits very high spatial resolution.

Conventional microwave measurement technique is limited to electron concentrations below $\sim 10^{13}/cm^3$ by the plasma cutoff frequency. The bridging of the gap between this plasma density range and densities

---

* Part 11 by F. C. Jahoda and G. A. Sawyer.

accessible to measurement with optical radiation remains a challenge to the extension of both technologies. It is in this spirit that we include some consideration of far infrared lasers as a natural extension of the optical domain.

In recent years some ultra-high density plasmas have also been produced, e.g., laser-produced sparks, in which the densities are so large that optical frequencies are not large compared to the electron plasma frequency. Refractive effects then no longer predominate compared to absorption and reflection; even when they are dominant, the phase shifts are so large that the techniques considered in this part (e.g., fringe counts), are mostly too fine a probe to cope with the required measurements. It is necessary, therefore, to state also an upper density limit, which we shall arbitrarily set here at an electron density times path length product of $\sim 10^{19}/cm^2$, equivalent to a phase shift of approximately $30 \times 2\pi$ for the ruby laser wavelength.

In Section 11.2, the refractive index is related to the dispersion relation, $\omega = \omega(k)$, and the dispersion relation for free electrons is derived for the case of small amplitude waves in nearly collisionless, cold plasma with or without a stationary magnetic field applied parallel to the direction of wave propagation. The validity of the approximations used is then discussed. Finally, the contributions of the nonelectron constituents of the plasma to the total refractivity are shown to be generally small.

In Section 11.3, we discuss various standard measurements of refractive effects. In particular, this covers interferograms, schlieren pictures, shadowgrams, and Faraday rotation measurements. By "standard measurements" we mean that the methods are well known with conventional light sources in aerodynamics, although these techniques did not become common for plasmas until the advent of the laser. The laser has the unique advantage of being a source sufficiently bright and monochromatic to compete favorably with the plasma self-luminosity in a narrow spectral band, while simultaneously being an effective collimated source on account of its small beam divergence. In addition, the giant pulses produced by Q-switching provide a convenient way of obtaining pictures of only a few nanoseconds exposure.

In Section 11.4, we discuss measurements of refraction phenomena by techniques that depend specifically on the long coherence lengths of laser light. These are (a) the coupled resonator gas laser interferometer technique for time resolved interferometry in which the laser is often both the source and phase sensitive detector and (b) holography, a new technique which can be adapted to make some of the standard measurements of Section 11.3, with particular experimental advantages.

A number of good reviews covering various portions of these topics already exist.[1-6] In addition, an extensive bibliography is given in Ref. 7.

11.2. Theory

11.2.1. Wave Equation

An electromagnetic (EM) wave exerts a force on the charged constituents (both bound and unbound) of any medium through which it propagates. This force gives rise to accelerations of the charges which in turn modify the time-varying electromagnetic field components of the wave, thereby altering its characteristics. Self-consistent solutions for an EM wave in a plasma can be obtained in principle by solving Maxwell's equations with appropriate boundary conditions together with the Lorentz force equation of motion for charged particles in an electromagnetic field. In general the magnetic forces exerted by a wave field are much weaker than the electric forces and we neglect the wave magnetic field in the treatment below. The results will be examined to determine that this neglect is justified.

In order to render the problem tractable, we make several simplifying assumptions. However, these do not detract from the utility of the solution for quite general situations. To begin with, we assume an infinite, homogeneous plasma and small amplitude waves. The small amplitude approximation permits the equations to be linearized, so that any sum of solutions is also a solution. It is then sufficient to discuss plane wave solutions, into which more general solutions can be Fourier-decomposed. We seek, then, plane wave solutions for the electric field propagating in the z direction of the form

$$\mathbf{E}(z, t) = \mathbf{E}_0 \exp[i(kz - \omega t)], \qquad (11.2.1)$$

---

[1] R. A. Alpher and D. R. White, *Pure Appl. Phys.* **21**, 431 (1965).

[2] Chapters by F. J. Weyl, J. W. Beams, R. Ladenburg, and D. Bershader, *in* "Physical Measurements in Gas Dynamics and Combustion," of "High Speed Aerodynamics and Jet Propulsion" (R. W. Ladenburg, ed.), Vol. IX. Princeton Univ. Press, Princeton, New Jersey, 1954.

[3] A. N. Zaidel, G. V. Ostrovskaia, and Yu. I. Ostrovskii, *Sov. Phys. Techn. Phys.* **13**, 1153 (1969).

[4] H. J. Kunze, *in* "Plasma Diagnostics" (W. Lochte-Holtgreven, ed.), p. 603. North-Holland Publ., Amsterdam, 1968.

[5] S. Martellucci, *Nuovo Cimento Suppl.* **5**, 642 (1967).

[6] U. Ascoli-Bartoli, *in* "Plasma Physics," p. 287. IAEA, Vienna, 1965.

[7] "Laser Applications in Plasma Physics," Bibliographical Ser. No. 35. IAEA, Vienna, 1969.

where it is understood that the complex notation is for mathematical convenience and the real part of **E** represents the physical quantity. The wave number $k = 2\pi/\lambda$ defines the space periodicity or wavelength λ and ω is the wave frequency in radians per second. Planes of constant phase are defined by the condition $kz - \omega t = $ const., from which the phase velocity v_{ph} is

$$v_{ph} \equiv dz/dt \,|_{\text{const. phase}} = \omega/k. \tag{11.2.2}$$

A solution requires specifying the "dispersion relation," the functional relation between ω and k. We turn now to this task, by first deriving a general wave equation for the electric field from the following three of Maxwell's equations:

$$\nabla \cdot \mathbf{E} = 4\pi\sigma, \tag{11.2.3}$$

$$\nabla \times \mathbf{E} = -\frac{1}{c}\frac{\partial \mathbf{B}}{\partial t}, \tag{11.2.4}$$

$$\nabla \times \mathbf{B} = \frac{1}{c}\left(4\pi\mathbf{j} + \frac{\partial \mathbf{E}}{\partial t}\right). \tag{11.2.5}$$

The units are Gaussian and σ and **j** refer, respectively, to total charge and current densities (both free and induced).

Taking the curl of both sides of (11.2.4), and substituting (11.2.5), we obtain

$$\nabla \times (\nabla \times \mathbf{E}) = -\frac{1}{c^2}\left(4\pi\frac{\partial \mathbf{j}}{\partial t} + \frac{\partial^2 \mathbf{E}}{\partial t^2}\right),$$

which, using the vector identity

$$\nabla \times (\nabla \times \mathbf{E}) \equiv \nabla(\nabla \cdot \mathbf{E}) - \nabla^2 \mathbf{E}$$

and (11.2.3) can be written as

$$\nabla^2 \mathbf{E} - 4\pi\nabla\sigma - \frac{4\pi}{c^2}\frac{\partial \mathbf{j}}{\partial t} - \frac{1}{c^2}\frac{\partial^2 \mathbf{E}}{\partial t^2} = 0. \tag{11.2.6}$$

This is the desired general wave equation.

For wave propagation in vacuum **j** and σ are both zero. In this case, (11.2.1) into (11.2.6) gives

$$k^2 = \omega^2/c^2 \tag{11.2.7}$$

or, from (11.2.2), the well known result that the phase propagation velocity of an electromagnetic wave in vacuum is the same for all frequencies: i.e., there is no dispersion.

Likewise substituting (11.2.1) into (11.2.3) shows that in vacuum ($\sigma = 0$) $\mathbf{k} \cdot \mathbf{E} = 0$. We retain the condition $\mathbf{k} \cdot \mathbf{E} = 0$ in the plasma, i.e., we consider only purely transverse waves. (This indeed corresponds to two distinct modes that can be sustained by a plasma when there is no static transverse magnetic field, as follows from a more general treatment.[8]) The wave equation then becomes

$$\nabla^2 \mathbf{E} - \frac{4\pi}{c^2}\frac{\partial \mathbf{j}}{\partial t} - \frac{1}{c^2}\frac{\partial^2 \mathbf{E}}{\partial t^2} = 0. \qquad (11.2.8)$$

Before it can be solved, it is necessary to determine the functional relation between $\mathbf{j}$ and $\mathbf{E}$. This will be done for 3 simple cases: (1) no static field and no collisions, (2) no static field but collisions included, and (3) static longitudinal $\mathbf{B}_0$ field but no collisions. We then content ourselves with citing more general results.

11.2.1.1. Solution for $B_0 = 0$, No Collisions. The simplest case to consider is a fully ionized plasma with no collisions and without an external magnetic field. The forces exerted on the particles by the electric field $\mathbf{E}$ of the wave are

$$m_e \, d\mathbf{v}_e/dt = -e\mathbf{E} \qquad (11.2.9)$$

and

$$m_i \, d\mathbf{v}_i/dt = Ze\mathbf{E} \qquad (11.2.10)$$

for electrons and ions of mass m_e and m_i, respectively, where Z is the ionic charge number and e the absolute value of the unit charge. The total current is $\mathbf{j} = e n_i \mathbf{v}_i - e n_e \mathbf{v}_e$, where n_i and n_e are the ion density and electron density, so that

$$\frac{\partial \mathbf{j}}{\partial t} = e\left(n_i \frac{\partial \mathbf{v}_i}{\partial t} - n_e \frac{\partial \mathbf{v}_e}{\partial t}\right). \qquad (11.2.11)$$

Using (11.2.9) and (11.2.10), this becomes

$$\frac{\partial \mathbf{j}}{\partial t} = e^2 \left(\frac{n_i Z}{m_i} + \frac{n_e}{m_e}\right) \mathbf{E}$$

$$\approx \frac{n_e e^2}{m_e} \mathbf{E} \qquad (11.2.12)$$

since $n_i Z = n_e$ and $m_i \gg m_e$.

Inserting (11.2.12) into the wave equation (11.2.8), we obtain

$$\nabla^2 \mathbf{E} - \frac{4\pi n_e e^2}{m_e c^2}\mathbf{E} - \frac{1}{c^2}\frac{\partial^2 \mathbf{E}}{\partial t^2} = 0. \qquad (11.2.13)$$

[8] L. Spitzer, Jr., "Physics of Fully Ionized Gases," p. 73. Wiley (Interscience), New York, 1956.

Again assuming a solution of the form (11.2.1), one obtains a dispersion relation

$$-k^2 - \frac{\omega_p^2}{c^2} + \frac{\omega^2}{c^2} = 0, \tag{11.2.14}$$

where ω_p, the plasma frequency, is defined by

$$\omega_p \equiv \left(\frac{4\pi n_e e^2}{m_e}\right)^{1/2}. \tag{11.2.15}$$

Rearranging (11.2.14) to read $k^2 = (\omega^2/c^2)(1 - \omega_p^2/\omega^2)$, comparing with (11.2.7), and using definition (11.2.2) gives the plasma refractive index, defined as the ratio of phase velocity in vacuum to phase velocity in the plasma:

$$\mu = \left(1 - \frac{\omega_p^2}{\omega^2}\right)^{1/2}. \tag{11.2.16}$$

11.2.1.2. Solution with Collisions Included. The effect of collisions between electrons and heavy particles is to randomize the electron velocity, or equivalently, to reduce the average electron velocity toward zero. This requires insertion of another term into the equation of motion (11.2.9):

$$\frac{d\mathbf{v}_e}{dt} = -\frac{e\mathbf{E}}{m_e} - \nu\mathbf{v}_e, \tag{11.2.17}$$

where ν is the effective number of collisions per second that a typical electron makes with heavy particles. As before, the ion equation can be neglected because of the large disparity between the electron and ion masses.

The solution $\mathbf{v}_e$ of (11.2.17) will have the same exponential time dependence as $\mathbf{E}$ so that (11.2.17) becomes $-i\omega\mathbf{v}_e = -e\mathbf{E}/m_e - \nu\mathbf{v}_e$. Therefore $\mathbf{v}_e = -e\mathbf{E}/m_e(\nu - i\omega)$ and

$$\frac{d\mathbf{j}_e}{dt} = -n_e e \frac{d\mathbf{v}_e}{dt} = \frac{n_e e^2}{m_e(\nu - i\omega)} d\mathbf{E}/dt$$

$$= \frac{n_e e^2 \mathbf{E}}{m_e(1 + i\nu/\omega)}. \tag{11.2.18}$$

Continuing as before with steps (11.2.12) through (11.2.15) gives the dispersion relation

$$k^2 = \frac{\omega^2}{c^2}\left(1 - \frac{\omega_p^2}{\omega^2(1 + i\nu/\omega)}\right). \tag{11.2.19}$$

The significance of k having a complex amplitude can be easily recognized as an attentuation of the wave as it progresses through the medium, which indeed would be expected in the presence of collisions, by noting that (11.2.1) becomes

$$\mathbf{E} = \mathbf{E}_0 \exp(-\text{Im}\{k\}z) \exp i(\text{Re}\{k\}z - \omega t).$$

Some simple algebra can quickly show that the imaginary part of (11.2.19) is positive definite and hence the real exponential term is always an attenuation. The expression for the phase velocity (11.2.2) must be generalized to $v_{ph} = \omega/\text{Re}\{k\}$.

11.2.1.3. Solution with Static Applied Magnetic Field.

Because an applied magnetic field exerts a force in a perpendicular direction on a charge moving with the velocity induced by the electric field of the wave, the equations of motion in two perpendicular Cartesian coordinates become coupled. There will exist, however, a favored coordinate system in which uncoupled equations result. Stated in other words, since a linear polarization can be decomposed into two oppositely rotating circular motions, one of which will correspond to the direction of the force on an electron in a magnetic field and the other of which will oppose the direction of this force, circular motions will be the more natural variables of the problem in this case.

In the absence of collisions, the equations of motion in a right-hand coordinate system for an electron subject to the electric field component of a transverse wave (which may have both x and y components) and a steady magnetic field, B_0, applied in the direction of wave propagation are

$$\frac{dv_x}{dt} = -\frac{e}{m_e}\left[E_x + \frac{v_y}{c}B_0\right],$$

$$\frac{dv_y}{dt} = -\frac{e}{m_e}\left[E_y - \frac{v_x}{c}B_0\right]. \qquad (11.2.20)$$

Inspection of these equations indicates that the linear combinations, $v_+ = v_x + iv_y$ and $v_- = v_x - iv_y$, reduce them to

$$dv_\pm/dt = -E_\pm e/m_e \pm i\omega_e v_\pm, \qquad (11.2.21)$$

where $\omega_e \equiv eB_0/m_e c$ (the electron gyrofrequency) and $E_\pm$ are defined by $E_\pm = E_x \pm iE_y$.

By our plane wave assumption, $E_\pm = \text{Re}\{E_{0\pm} \exp i(k_\pm z - \omega t)\}$, and therefore the effect of the $\pm i$ factor ($e^{\pm i\pi/2}$) of the y component is to advance or retard its phase by 90° relative to the x component. As E_x and E_y oscillate in time, the two-dimensional vectors $E_\pm$ will rotate in the xy plane. For the special case of $|E_x| = |E_y|$, two oppositely rotating "circular polarizations" result.

The program now is to solve for $k_\pm$ as functions of ω with the general wave equation (11.2.8) (the current vector will differ for $\mathbf{E}_+$ and $\mathbf{E}_-$, leading to two solutions) and then solve for E_x and E_y (real parts of the oscillatory functions) from the definitions of E_+ and E_-.

Since $v_\pm$ will have the same time dependence as $E_\pm$, $dv_\pm/dt = -i\omega v_\pm$ and the solution of (11.2.21) is

$$v_\pm = -i\frac{e}{m_e}\frac{E_\pm}{(\omega \pm \omega_e)}. \qquad (11.2.22)$$

The (complex) current due to electrons is

$$j_\pm = -en_e v_\pm$$

and the electron term to be substituted into the second term of (11.2.8) becomes

$$-\frac{4\pi}{c^2}\frac{\partial j_e}{\partial t} = -\frac{4\pi}{c^2}(-en_e)\left(-i\frac{e}{m_e}\frac{\partial E_\pm/\partial t}{(\omega \pm \omega_e)}\right)$$

$$= -\frac{\omega_p^2}{c^2}\left(\frac{\omega}{\omega \pm \omega_e}\right)E_\pm. \qquad (11.2.23)$$

For the ions, carrying through the identical derivation (11.2.20)–(11.2.23), with due regard for the change in mass and charge polarity, gives

$$-\frac{4\pi}{c^2}\frac{\partial j_i}{\partial t} = -\frac{\omega_p^2}{c^2}\frac{m_e}{m_i}\frac{\omega}{(\omega \mp \omega_i)}E_\pm, \qquad (11.2.24)$$

where $\omega_i \equiv eB_0/m_i c$.

With our assumption, $\omega \gg \omega_i$, the possible resonance condition of (11.2.24) ($\omega \approx \omega_i$) is avoided, and because of the factor m_e/m_i it can be neglected in comparison to (11.2.23). The wave equation (11.2.8) then becomes

$$\nabla^2 E_\pm - \frac{\omega_p^2}{c^2}\left(\frac{\omega}{\omega \pm \omega_e}\right)E_\pm - \frac{1}{c^2}\frac{\partial^2 E_\pm}{\partial t^2} = 0, \qquad (11.2.25)$$

which, substituting (11.2.1), yields the dispersion relation

$$-k_\pm^2 - \frac{\omega_p^2}{c^2}\frac{1}{(1 \pm \omega_e/\omega)} + \frac{\omega^2}{c^2} = 0$$

or

$$k_\pm = \frac{\omega}{c}\left[1 - \frac{\omega_p^2}{\omega^2(1 \pm \omega_e/\omega)}\right]^{1/2}. \qquad (11.2.26)$$

The solution for E_x and E_y in terms of E_+ and E_- will be deferred to Section 11.3.3 on Faraday rotation, where we show that the physical consequence of (11.2.26) is that a linearly polarized wave has its plane of polarization rotated when propagating along a magnetic field.

For an applied magnetic field transverse to the direction of propagation, two cases can be distinguished. Either the applied B field is parallel to the wave E field, in which case it has no influence on the particle motions induced by the wave and is equivalent to the case of no field, or it has a component perpendicular to the E field, in which case it induces z currents, and mixes the transverse wave modes with a longitudinal mode. We need not consider the complexities of this case, and state only in passing that the two roots of the dispersion relation have smaller correction factors due to the magnetic field ($\sim \omega^{-2}$ rather than ω^{-1} as in the $B \parallel k$ case) and do not involve the circular polarization components introduced above. Therefore, for the general case of arbitrary applied B field direction the rotation of the plane of polarization of a transverse wave depends only on the longitudinal field component.

11.2.2. Discussion of Dispersion Relations and Assumptions

The three cases above are special cases of a single dispersion relation:

$$k^2 = \frac{\omega^2}{c^2}\left[1 - \frac{\omega_p^2}{\omega^2(1 \pm \omega_e/\omega)[1 + i(v/\omega)]}\right]. \quad (11.2.27)$$

The refractive index is then

$$\mu = \left[1 - \frac{\omega_p^2}{\omega^2(1 \pm \omega_e/\omega)[1 + i(v/\omega)]}\right]^{1/2}$$

which, when all frequencies are small compared to ω, can be expanded as

$$\mu = 1 - \frac{1}{2}\frac{\omega_p^2}{\omega^2}\left(1 \mp \frac{\omega_e}{\omega}\right)\left(1 - i\frac{v}{\omega}\right). \quad (11.2.28)$$

The limits imposed by the requirements $\omega_p, \omega_e, v \ll \omega$ can be restated in terms of density, magnetic field, and temperature as follows (in cgs units): For the ruby laser wavelength (6943 Å),

$$\omega = 2\pi c/\lambda = 2.7 \times 10^{15}/\text{sec}.$$

The electron plasma frequency is

$$\omega_p = (4\pi n_e e^2/m_e)^{1/2} = 5.6 \times 10^4 n_e^{1/2}/\text{sec}.$$

The electron cyclotron frequency is

$$\omega_e = eB_0/m_e c = 1.8 \times 10^7 B_0/\text{sec}.$$

For the effective collision frequency v we cannot be quite so precise. Taking the relaxation time, t_{ei}, for 90° total deflection of electrons as a result of distant collisions with singly charged ions at a common temperature T, one obtains[9]

$$v = \frac{1}{t_{ei}} = \frac{4\pi e^4 n_i \ln \Lambda}{(2m_e)^{1/2}(\tfrac{3}{2}kT)^{3/2}},$$

where $\ln \Lambda$ is a slowly varying parameter of order 10. For a hydrogen or deuterium plasma $n_i = n_e$ and

$$v \approx 4.3 \times 10^{-5} n_e/(kT)_{ev}^{3/2}/\text{sec}.$$

Thus, arbitrarily requiring all of these frequencies to be less than 0.1ω for the validity of (11.2.28) one obtains

$$n_e < 2.3 \times 10^{19}/\text{cm}^3, \qquad B < 1.5 \times 10^7 \text{ G}$$

and, even assuming the extreme allowed value of n_e, kT need only be greater than ~ 2.5 eV.

For the lower frequencies of the infrared gas lasers considered later in this part, the upper density limit decreases as the square of the ratio of the infrared laser wavelength to the ruby laser wavelength. Since, however, these lasers are used only to extend the lowest density limit attainable with refractivity measurements, the criterion that the plasma frequency be small compared to the wave frequency is very well satisfied in all cases considered.

It has been assumed that the force due to the wave magnetic field on the electrons is small compared to the force due to the electric field. This requires $(v/c)B_{\text{wave}} \ll E_{\text{wave}}$. The magnitude of B and E, however, are related by the Maxwell equation (11.2.4), which, for our transverse plane wave becomes $kE = (\omega/c)B$. Therefore $(v/c)B$ will be much less than E if $v \ll \omega/k$, i.e., if the particle velocity is small compared to the wave phase velocity. This is always the case for transverse EM waves, for which, as given by (11.2.28), the phase velocity exceeds the velocity of light. Indeed we have neglected all particle velocities not induced by the wave, the so-called cold plasma approximation. For nonrelativistic plasmas, the finite temperature correction to the refractivity[5] is only of order $kT_e/m_e c^2$ and can be safely ignored.

The small-amplitude-wave approximation is implicit throughout our derivation, primarily in the linearization which permits the use of complex exponentials and the neglect of higher-order terms in the equations of

---

[9] S. Glasstone and R. H. Lovberg, "Controlled Thermonuclear Reactions," p. 97. Van Nostrand, Princeton, New Jersey, 1960.

motion. References to more sophisticated treatments of the dispersion relations are given by Martellucci.[5]

That the refractive index is less than unity, indicating a phase velocity greater than c, is characteristic of free electrons and not in conflict with special relativity. Any real signal contains a spread of frequencies and its center of disturbance (as distinct from the various phase velocities of its frequency components) travels with the group velocity $v_g = d\omega/dk$, which is always less than c. Equivalently, the establishment of a "pure" oscillation requires time for the initial transients to decay, the time becoming longer the more nearly monochromatic the oscillation. The concept of phase velocity really implies only a comparison of phase differences between two given points in space with and without the medium present after sufficient time has elapsed to establish steady state conditions. Refractivity measurements in monochromatic light from observed fringe displacements determine the phase velocity. In white light interferometry the center of contrast within the fringe pattern, however, depends on the group velocity, and, particularly when free electrons dominate, it will not coincide with the displacement of individual fringes. Further references and an illustration of this effect are presented by Alpher and White.[1]

11.2.3. Nonelectronic Refractivity Contributions

So far we have considered mainly the refractivity of the free electron component of the plasma. The refractivity of completely stripped ions has been shown to be negligible in comparison to that of electrons because of the large ion to electron mass ratio. If there are other plasma components, e.g., atoms, molecules, or incompletely stripped ions, their contribution to a weighted sum of terms for the total refractivity must also be considered. For highly ionized plasmas these will be minority constituents and knowledge that their specific refractivity (i.e., contribution per particle) is not large compared to that of the electron suffices to justify neglecting their contribution merely on the basis of their relative number. This is a fortunate circumstance since accurate determination of specific refractivities is both theoretically and experimentally a complex task which must be considered individually not only for every different atom and ion but also for every possible excitation state. While the subject is treated by a vast literature, it is still very incomplete.

For bound electrons in the ground state of an atom, a dispersion relation of the form

$$\mu = 1 + \frac{2\pi e^2 n_e}{m_e} \sum_k \frac{f_k}{\omega_k^2 - \omega^2} \qquad (11.2.29)$$

can be derived by the same general procedure used for deriving the relation for free electrons. Here, f_k is the oscillator strength for transitions between the ground state and a discrete energy level k, and ω_k is the frequency difference in radians per second between these levels. Given that oscillator strengths are of order unity or less and that the strong resonance transitions are generally at higher frequencies than the optical range, it follows that the absolute value of the deviation from unity for the refractivity of these bound charges is smaller than that of free electrons provided the frequency ω is sufficiently far from any ω_k. When $\omega \sim \omega_k$, the formulation leading to (11.2.29) breaks down because dissipation has been neglected. These are the regions of high "anomalous" refractivity. They are avoided by choosing ω well away from any resonance spectral line of the species under consideration.

The refractivity of incompletely stripped ions in the ground state as deduced from various calculations of polarizability[10] is of the order of magnitude, but systematically less than that of the neutral atoms.[5] While the refractivity contributions of excited states of atoms and ions scale approximately with the fourth power of the mean distance of the excited electron from the nucleus, their significantly smaller concentration again limits the total effect.

When $\omega \ll \omega_k$, Eq. (11.2.29) can be approximated by the Cauchy formula

$$\mu - 1 = A + B/\lambda^2; \qquad (11.2.30)$$

the empirically determined coefficients A and B can be found in handbook references for several gases.

As an indication of the magnitude of the various refractivities we compare the contribution of H_2 molecules and H atoms with that of electrons at the ruby laser wavelength:

$$\frac{(\mu - 1)_{\text{electrons}}}{(\mu - 1)_{H_2}} = 42 \quad \text{and} \quad \frac{(\mu - 1)_{\text{electrons}}}{(\mu - 1)_{H}} = 51.$$

In determining these ratios, we have calculated the electron refractivity from (11.2.28), used an experimental value for the hydrogen molecule refractivity, and derived the hydrogen atom value from a low frequency value of the polarizability[11] which gives a good approximation to the more accurate value that would be obtained if empirical constants A and B for hydrogen atoms were available.

---

[10] R. A. Alpher and D. R. White, *Phys. Fluids* 2, 153 (1959).

[11] C. W. Allen, "Astrophysical Quantities," 2nd ed. Oxford Univ. Press, London and New York, 1963.

Since A is always very much larger than B in (11.2.30), the change with wavelength of the refractivity when a Cauchy relation holds is very much less than that of the electron contribution. The departure from unity of the electron contribution varies as λ^2, which follows from (11.2.28).

Thus in cases of doubt or when additional sources of refractivity cannot be ignored, the electron contribution alone can be singled out if measurements of the refractivity are made at two different wavelengths.[12,13]

11.3. Standard Experimental Methods

11.3.1. Optical Interferometry

Optical interferometry is a well established technique for measuring the refractivity of transparent materials, by directly comparing the phase of the wavefront of interest with that of a reference wavefront. It is only in the last ten years that the applicability of interferometry to measuring densities of ionized plasmas has been recognized. In these few years, however, the method has been used extensively and several good reviews have appeared.[1,5,6]

11.3.1.1. *The Mach–Zehnder Interferometer.* Several types of two-beam interferometers have been devised but the most widely used for plasma measurements is the Mach–Zehnder interferometer[14] (Fig. 1). A unique

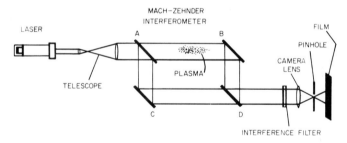

Fig. 1. Diagram of Mach–Zehnder interferometer.

property of the Mach–Zehnder geometry is that the fringes can be localized at will. This feature is often desirable in plasma work, because the fringe pattern can be localized in the plane of the plasma, so that the fringe pattern and the plasma, or some object immersed in the plasma, can be photographed together. This capability is desirable, for example, when it

---

[12] R. A. Alpher and D. R. White, *Phys. Fluids* **2**, 162 (1959).
[13] U. Ascoli-Bartoli, A. DeAngelis, and S. Martellucci, *Nuovo Cimento* **18**, 1116 (1960).
[14] W. Kinder, *Optik (Stuttgart)* **1**, 413 (1946).

is necessary to see the positions of electrodes in the plasma. Other features of the Mach–Zehnder interferometer are: (a) the light makes only one pass through the test volume in contrast to, for example, the Michelson interferometer where two passes are made through the test volume, and (b) the test and comparison beams are widely separated in space, which is convenient in avoiding obstructing apparatus.

The Mach–Zehnder interferometer has four mirrors (Fig. 1). Mirrors A and D are half-silvered beam splitting mirrors, and mirrors B and C are totally reflecting. The interference fringe pattern on the screen is determined by optical path difference between the two arms of the interferometer.

It is convenient to adjust the interferometer, by tilting at least one of the mirrors of Fig. 1, so that there is a slight angle between the recombined wavefronts. This produces several straight fringes in the field when there is no plasma and provides a background pattern against which the plasma effect appears. The background allows one to determine whether the plasma changes are in the direction of increasing or decreasing density, based on the direction of displacement of the background fringes. The background fringe spacing also provides a calibration against which fractional fringe shifts can be estimated.

If one neglects the refractive effects due to changes in the neutral gas density in the path, and this is usually valid, the effective length change or shift due to the plasma in units of fringes is

$$N = \int [(\mu - 1)/\lambda] \, dl = \frac{-1}{2\pi} \frac{e^2 \lambda}{mc^2} \int n_e \, dl$$

$$= \frac{-1}{3.2 \times 10^{17}} \int n_e \, dl \quad \text{for} \quad \lambda = 6943 \, \text{Å},$$

(11.3.1)

where the electron index of refraction μ is taken from Eq. (11.2.28) with $\nu = 0$ and $\omega_e = 0$, the plasma frequency is given by Eq. (11.2.15), and $\omega = 2\pi c/\lambda$ has been substituted.

Figure 1 shows a typical example of interferometric measurement of the plasma density in a theta pinch. The Mach–Zehnder interferometer is so arranged that the light beam travels longitudinally through the plasma. The reference beam travels through the compensating chamber below. In the example shown, the interferometer is illuminated by a ruby laser whose beam is expanded with a telescope. The camera incorporates an interference filter that passes only a narrow band around the laser wavelength in order to eliminate plasma light at other wavelengths, and a pinhole to further reduce unwanted light which is not parallel to the laser beam.

Any interferometer requires precision adjustment devices such as micrometer screws on the mirror mounts to allow fine adjustment of the mirrors, and high quality optical flats for the mirrors and for the windows on the plasma chamber. The window flatness must match the accuracy desired in the fringe measurement; i.e., to measure fringe shifts to $\lambda/10$ accuracy requires windows for which the deviation from flatness is less than $\lambda/10$.

Figure 2 shows a plasma interferogram taken with the apparatus of Fig. 1. The picture is a contour map showing contours of constant integrated plasma density $\int n_e \, dl$. The low-density plasma shows a shift of about 6 fringes, corresponding to an integrated density of $1.9 \times 10^{18}/\text{cm}^2$.

11.3.1.1.1. Fringe Localization. If the plasma to be measured is located midway between mirrors A and B as shown in Fig. 1, it is desirable that the arms AC and BD be exactly half as long as arms AB and CD. This locates mirror C at the same distance through the optical train as the plasma, and thus gives mirror C a virtual position in the plasma.[14] If the interferometer has been adjusted to localize the fringes in the plane of the plasma, the fringes are also localized at mirror C. The number of fringes in the background field can then be adjusted by tilting mirror C without affecting the fringe localization. This is a considerable practical advantage. Tilting any of the other mirrors would affect both localization and the number of fringes in the field of view.

11.3.1.1.2. Light Sources. Both conventional and laser light sources can be used for interferometry. High-pressure mercury lamps filtered to a single spectral line are often used.[15] The light must be collimated into a parallel beam. In a conventional light source, this is achieved by using a pinhole aperture and a collimating lens; in consequence, high intensity is required in the source. In a laser source the light beam is already collimated but is of small diameter and in this case a telescope can be used to blow up the laser beam to any desired diameter as shown in Fig. 1. This has the further advantage of reducing the already small laser beam divergence by the ratio of the focal lengths of the telescope lenses. The common focal point within the telescope is equivalent to the pinhole aperture of a conventional source. The light must be collimated well enough so that the diverging rays passing through any object point are less than a $\frac{1}{4}$-wave

---

[15] E. Fünfer, K. Hain, H. Herold, P. Igenbergs, and F. P. Küpper, *Z. Naturforsch.* **A17**, 967 (1962).

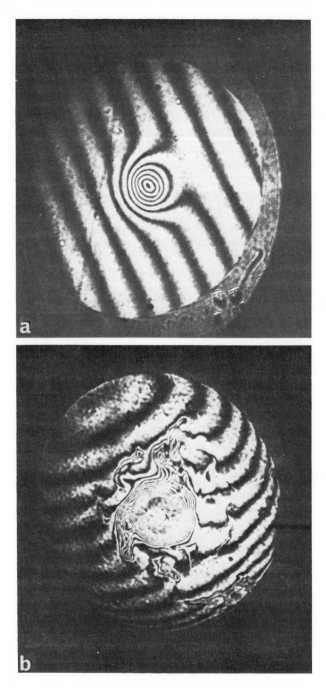

Fig. 2. Interferograms of Scylla IV θ-pinch plasma. (a) Stable low-density plasma; (b) unstable, very-high-density plasma.

out of phase due to the angular divergence. The condition on divergence, θ, is

$$\left(\frac{h}{\cos \theta/2} - h\right) < \frac{\lambda}{4}, \tag{11.3.2}$$

where h is the distance between source and collimation lens.

The same relation places a restriction on the allowable path difference between the plasma and reference beams. In this case h refers to the mean path difference between the two beams.[14] The more restrictive case applies.

The permissible spectral width of the source depends on the largest order number of interference that must be detectable. The fringe contrast will disappear due to overlapping orders of different wavelength light when

$$N\lambda_1 = (N + \tfrac{1}{2})\lambda_2$$

or (11.3.3)

$$N \sim \lambda/(2\,\Delta\lambda).$$

Thus for a source wavelength spread of 1 Å at 5460 Å (the mercury green line), the last fringe visible on either side of the central fringe is $N = 2730$. This also places an upper limit on the inequality of path lengths in the two arms of the interferometer. When the path lengths become different enough so that the fringes in the image plane fall outside the range allowed by Eq. (11.3.3), overlapping of different wavelength fringes will destroy the contrast. Thus in the example above, $N = 2730$ at 5460 Å corresponds to a path length $N\lambda = 0.15$ cm, and this is the maximum allowed path difference in the two arms of the interferometer. Finally, the light source must be bright enough that the interference pattern dominates whatever plasma self luminosity reaches the film plane.

Even though conventional light sources are satisfactory, laser light sources are much better, and they have made interferometry of plasmas easier. The laser source is sufficiently monochromatic and has small enough divergence that fringes can be achieved even with a difference of several centimeters between the two beams of the interferometer. This means that interferometer adjustment is much less exacting than with a conventional source. Also, the laser is so bright that it is always easy to overcome plasma luminosity. In addition, a giant pulse ruby laser provides a conveniently short pulse of light when good time resolution is needed. If, however, it is desired to take streak or framing camera photographs of the fringes over an extended time period the short pulse of ruby laser light is generally unsuitable and a conventional light source is usually preferred.[15,16] Nevertheless, by cooling a free-running ruby laser, a continuous output has been achieved for 100 μsec, enabling the use of standard framing camera techniques.[17]

11.3.1.1.3. EFFECT OF BENDING OF THE RAYS. If there are appreciable density gradients in the plasma, transverse to the direction of light propagation, they will produce deflection of the light rays in the plasma. (See Section 11.3.2.) The apparent localization of the fringe pattern can be changed or, if the bending is severe, overlapping of fringes can wash out the pattern just as source divergence does. But the most important effect is usually a spurious variation of the optical path length caused by the curvature of the deviated ray.

In order to determine when this effect is significant it is useful to consider the lowest order correction for a uniform linear variation of the refractive index in the transverse direction.* Consider ray AB in Fig. 3.

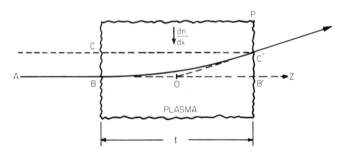

FIG. 3. Effect of ray bending on path length.

As it moves through the plasma of length l it will bend as indicated and emerge at C′ rather than B′. By imaging with the camera lens of the interferometer on the exit plane P we actually determine the phase shift of ray BC′ instead of the assumed path CC′ that ignores the refractive deviation. This error in our estimate of the phase shift along line CC′ is due, therefore, to the optical path length difference along BC′ compared with that along CC′. The optical path difference in units of wavelength is

$$\delta = \frac{1}{\lambda}\int_B^{C'} \mu_{BC'}\, ds - \frac{1}{\lambda}\int_C^{C'} \mu_{CC'}\, dz \qquad (11.3.4)$$

[16] R. D. Medford, A. L. T. Powell, A. G. Hunt, and J. K. Wright, *Proc. Int. Conf. Ioniz. Phenomena Gases, 5th,* 1961, **2**, p. 2000. North-Holland Publ., Amsterdam, 1962
[17] E. M. Little, W. E. Quinn, and G. A. Sawyer, *Phys. Fluids* **8**, 1168 (1965).
[18] R. E. Siemon, Los Alamos Sci. Lab., Los Alamos, New Mexico, 1969.

* We are indebted to R. E. Siemon[18] for help with this formulation.

where $ds = (dx^2 + dz^2)^{1/2} = dz(1 + (dx/dz)^2)^{1/2}$. The differential dx/dz is just the angular deflection α, which will be shown in Section 11.3.2.2 to be

$$\alpha = dx/dz \approx (\partial \mu/\partial x) z \quad \text{for} \quad \mu \approx 1. \tag{11.3.5}$$

Thus $ds = dz(1 + (\partial\mu/\partial x)^2 z^2)^{1/2} = dz(1 + \tfrac{1}{2}(\partial\mu/\partial x)^2 z^2)$ since $\partial\mu/\partial x$ is of the order of the refractive index decrement from unity and $(\partial\mu/\partial x)^2 z^2 \ll 1$.

Integrating (11.3.5) for $\partial\mu/\partial x = $ constant gives

$$x = \tfrac{1}{2} (\partial\mu/\partial x) z^2 \tag{11.3.6}$$

for the path BC'. Thus

$$\mu_{CC'} = \mu_{BB'} + (\partial\mu/\partial x)\,\Delta x\,|_{z=l} = \mu_{BB'} + \frac{1}{2}\left(\frac{\partial\mu}{\partial x}\right)^2 l^2,$$

$$\mu_{BC'} = \mu_{BB'} + \frac{\partial\mu}{\partial x}\left(\frac{1}{2}\frac{\partial\mu}{\partial x}z^2\right),$$

and

$$\lambda\delta = \int_{z=0}^{l} \left(\mu_{BB'} + \frac{1}{2}\left(\frac{\partial\mu}{\partial x}\right)^2 z^2\right)\left(1 + \frac{1}{2}\left(\frac{\partial\mu}{\partial x}\right)^2 z^2\right) dz$$

$$- \int_{z=0}^{l} \left(\mu_{BB'} + \frac{1}{2}\left(\frac{\partial\mu}{\partial x}\right)^2 l^2\right) dz$$

$$= \frac{1}{6}\left(\frac{\partial\mu}{\partial x}\right)^2 l^3 + \frac{\mu_{BB'}}{6}\left(\frac{\partial\mu}{\partial x}\right)^2 l^3 - \frac{1}{2}\left(\frac{\partial\mu}{\partial x}\right)^2 l^3$$

$$\simeq -\frac{1}{6}\left(\frac{\partial\mu}{\partial x}\right)^2 l^3, \tag{11.3.7}$$

where in the last step we have again used $\mu_{BB'} \approx 1$ and the fact that $\tfrac{1}{2}(\partial\mu/\partial x)^2 z^2 \ll 1$ to drop the higher order term.

Finally, we can assess the fractional error due to refractive bending as δ/N, where N is the observed fringe displacement compared to vacuum and assumed to depend on the refractive index along CC' as (11.3.1) $N = (\mu_{CC} - 1)l/\lambda$:

$$\frac{\delta}{N} = -\frac{(\partial\mu/\partial x)^2 l^2}{(\mu - 1)6} \tag{11.3.8}$$

From (11.3.1) we have

$$\mu - 1 = \frac{-1}{2\pi}\frac{e^2}{mc^2}\lambda^2 n_e,$$

$$\frac{\partial\mu}{\partial x} = \frac{-1}{2\pi}\frac{e^2}{mc^2}\lambda^2 \frac{\partial n_e}{\partial x},$$

and

$$\frac{\delta}{N} = \frac{l^2}{6}\frac{e^2\lambda^2}{2\pi mc^2}\frac{(\partial n_e/\partial x)^2}{n_e}. \tag{11.3.9}$$

As a numerical example let $n_e \approx 10^{17}/\text{cm}^3$ and $\partial n_e/\partial x = 10^{17}/\text{cm}^4$, $\lambda = 6.9 \times 10^{-5}$ cm (ruby), and $l = 100$ cm. Then (11.3.9) gives $\delta/N = 0.035$.

Since the error increases with the square of the length, for long plasma devices the correction can be significant and may need to be taken into account. It may then be necessary to use the eikonal equation[19] which deals directly with the phase of the emergent wavefront. Furthermore, if the angular deviation approaches the limit put on the source divergence (see Section 11.3.1.1.2), the fringes will wash out.

Since BC' is a parabola, the emergent ray will project back to the axis at the center of the plasma. If the plasma midplane instead of the exit plane is imaged, the refractive bending error is only half that given above. This follows from a similar derivation comparing path BC' to BOC' and remembering that the imaging optics exactly compensate for the geometrical path difference between OC' and OB'.

11.3.2. Schlieren Methods

11.3.2.1. Introduction. The bending of light rays by transverse density gradients, which was introduced in the previous section as a limitation on the utility of interferometry, is itself the basis of a considerable variety of diagnostic procedures to detect spatial variations of the refractive index. It is now customary to designate all such methods as "schlieren," a German word meaning striations in transparent solids, although the older literature reserves the term schlieren for techniques that detect the uniform deflection of a constant refractive index gradient and uses the term shadowgraph for techniques used to observe nonuniform deflections due to varying gradients.

The practical distinction between them is whether or not the detection occurs at the location of the focused image of the refractive disturbance. In the image plane all rays from a given object point, irrespective of their angular divergence, are recombined in a common image point (using an ideal lens), and there will be no intensity fluctuations due to refractive bending. A schlieren system introduces subsidiary elements into the optical system (e.g., a knife edge, as considered below) that do cause intensity variations in the location of the focused image. A shadowgraph system, on the other hand, examines the intensity variations in a plane that does not contain the location of the focused image; usually a collimated light beam is used without imaging lenses. With the shadowgraph system and constant density gradients one can obtain only an outline of the

---

[19] M. Born and E. Wolf, "Principles of Optics," p. 111. Pergamon, Oxford, 1959.

boundary within which the constant gradient exists, whereas if the gradients vary within these regions there may also be a richness of detail in intensity fluctuations due to variable amount of overlap of initially parallel rays. Both schlieren and shadowgraph methods can be very effective in delineating discontinuities and, in general, indicating overall plasma configurations; but in both cases quantitative evaluation is difficult and generally requires further specialized arrangements.

11.3.2.2. Derivation of Angular Deflection.[20] For the case of a constant density gradient perpendicular to the direction of propagation, the amount of angular deflection is easily derived with the aid of Fig. 4. A section

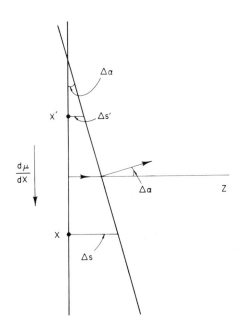

FIG. 4. Angular deviation of wavefront due to density gradient.

of wavefront propagating along z arrives simultaneously at x and x', which demarcate the incident boundary of the region containing the gradient. In a subsequent time interval Δt, the portions incident at x and x' travel distances proportional to their velocities, $\Delta s = c\,\Delta t/\mu$ and $\Delta s' = c\,\Delta t/\mu'$, respectively, where the refractive index μ' at x' is related to the refractive index μ at x by $\mu' = \mu + (\partial\mu/\partial x)\,\Delta x$ and $\Delta x = x - x'$. Thus

[20] H. Wolter, in "Handbuch der Physik" (S. Flügge, ed.), Vol. 24, p. 555. Springer, Berlin, 1956.

and
$$\Delta s' = \Delta s(\mu/\mu')$$
$$\Delta s - \Delta s' = \Delta s(1 - \mu/\mu') = (\Delta s/\mu)(\partial \mu/\partial x)\, \Delta x. \quad (11.3.10)$$

Moreover, the line joining the newly advanced positions at $\tau = \Delta t$ represents the advancing wavefront whose normal now deviates from the z direction by the angle $\Delta\alpha$. Projecting the new wavefront until it intersects the x axis also at the angle $\Delta\alpha$ (at a point which we may take as the zero of coordinates) one can also express the difference $\Delta s - \Delta s'$ in terms of $\Delta\alpha$. Namely, $\Delta s = x\Delta\alpha$ and $\Delta s' = x'\Delta\alpha$, so that

$$\Delta s - \Delta s' = \Delta\alpha(x - x') = \Delta\alpha\, \Delta x \quad (11.3.11)$$

and, combining with Eq. (11.3.10), we obtain the desired result

$$\Delta\alpha = (1/\mu')(\partial\mu/\partial x)\, \Delta s. \quad (11.3.12)$$

For a nonconstant density gradient the derivation can be generalized by a limiting process to $d\alpha = (1/\mu)(\partial\mu/\partial x)\, ds$, and as long as the total deflection from the z axis remains small, we obtain

$$\alpha_x = \int_0^z (1/\mu)(\partial\mu/\partial x)\, dz$$

and of course equivalently,

$$\alpha_y = \int_0^z (1/\mu)(\partial\mu/\partial y)\, dz.$$

For a plasma, the value of refractive index μ in the denominator can be taken as unity to a very good approximation, and evaluating the derivative from the formula for the electron index of refraction given by Eq. (11.2.16) we find the deflection angle in radians to be

$$\alpha = -(\lambda^2/2\pi)(e^2/m_e c^2) \int (dn_e/dx)\, dz \quad (11.3.13a)$$

which in cgs units, using the average value of the density gradient over the integration length Δl, becomes for the ruby wavelength

$$\alpha = -2.1 \times 10^{-22}(\Delta n_e/\Delta x)\, \Delta l. \quad (11.3.13b)$$

The significance of the minus sign is that the deflection is in the direction of decreasing electron density.

11.3.2.3. Detection of Angular Deflection

11.3.2.3.1. FOCUSED IMAGE (SCHLIEREN). The simplest kind of schlieren system is illustrated in Fig. 5. The first lens L_1 forms a collimated beam from the point source S. A refractive disturbance, represented by the

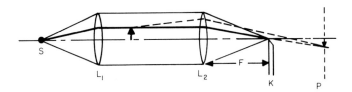

FIG. 5. Diagram of schlieren apparatus.

arrow head, deviates the ray through this point along the dotted path instead of the solid line path it would take in the absence of a disturbance. In either case, as already stated above, the ray passes through the same location in the plane P in which lens L_2 images the disturbance. In the focal plane of lens L_2, however, every incident direction θ has a unique spatial distance from the optical axis equal to $F\theta$ where F is the focal length of L_2. Thus, the undeviated ray and all other undeviated rays cross the axis in the focal plane (this point is an image of the point source), whereas the dotted ray passes above the axis. A knife edge K that just obstructs the axis will, in the absence of any refractive disturbance, prevent all source light from reaching the image plane P, whereas in the presence of refractive bending all rays bent upward from the axis will pass the knife edge and contribute to an image at P of the region of disturbance. In actual practice the source S has a finite extent, and if its image is cut by the knife edge, there will be some intermediate uniform illumination of the image plane in the absence of refractive disturbances, with brightening or darkening by refractive bending according to whether the angular deflections bend the light into the free or the obstructed half plane. Such a system responds only to the component of the gradient perpendicular to the knife-edge orientation, since displacements along the knife edge do not affect the image intensity.

In order to get good sensitivity, i.e., large fractional intensity changes as well as linearity, it is desirable to have the source image (as seen past the knife edge) be a slit.[21] This is readily accomplished by covering approximately half the actual source area with a second knife edge in such a manner that its image in the focal plane of lens L_2 in combination with the knife edge actually located there forms a virtual slit. This procedure, however, is wasteful of the luminous flux available from the source. Techniques with spark light sources[21,22] have been improved with the use of giant pulse ruby lasers.[23,24] Perhaps the ultimate in laser sources

[21] R. H. Lovberg, *IEEE Trans. Nucl. Sci.* **11,** 187 (1964).
[22] U. Ascoli-Bartoli and S. Martellucci, *Nuovo Cimento* **27,** 475 (1963).
[23] U. Ascoli-Bartoli, S. Martellucci, and E. Mazzucato, *Nuovo Cimento* **32,** 298 (1964).
[24] R. H. Lovberg, *AIAA J.* **4,** 1215 (1966).

has been reached in the subnanosecond resolution, multiple frame schlieren photography of a laser-induced gas breakdown achieved by Alcock et al.,[25] with the second-harmonic radiation of a mode-locked neodymium-glass laser.

A simple quantitative measurement with time resolution, but only for a single ray through the plasma, has been achieved with a gas laser source and photomultiplier detection on a shock tube experiment.[26] A gas laser beam, 1.25-mm diameter, is passed through the plasma disturbance and is then incident on a knife edge set to half obscure the beam. A photomultiplier behind the knife edge detects changes in the intensity due to bending of the beam. Quantitative measurement is greatly simplified in this single ray probing of the plasma.

11.3.2.3.2. DOUBLE-INCLINED-SLIT SCHLIEREN. Quantitative evaluation retaining spatial resolution can be greatly improved with the double-inclined-slit method.[24] The source must be in the form of a slit S_1, assumed to be vertical, and a second slit S_2, inclined to the first, replaces the knife edge of Fig. 5. The undeviated rays get through the crossed-slit combination only for the one "point" of the source slit image located at the geometrical intersection of this image and the inclined slit S_2. All the other parts of the image of the source slit S_1 formed by undeviated rays are blocked since they will miss the open aperture of the inclined slit S_2. Now consider a particular small region of the plasma in which a refractive index gradient will deviate rays through an angle β in the horizontal direction. All parts of the source slit S_1 contribute some to the illumination of this small region, and the deviated rays will form an image S_1' of the source slit displaced laterally a distance $F\beta$ from the image of S_1 in undeviated light. Again only one "point" of the (displaced) source image will pass through the inclined slit aperture. This single point image of S_1 is formed at the point on slit S_2 where it is separated by a distance $F\beta$ from the undeviated image of S_1.

Thus, a sorting of the rays, correlating angular deviation with position in the plane of the intersecting slits is accomplished. Since the apparent slit intersection moves along the inclined slit with increasing deflection either a horizontal or vertical component deflection may be singled out, although its origin for a vertical source slit remains a horizontal plasma deviation. A cylindrical lens with its axis horizontal is placed to image the crossed-slit plane on the plane P, where, in the absence of the cylindrical lens the plasma is imaged. The cylindrical lens leaves the horizontal dimension of

[25] A. J. Alcock, C. DeMichelis, and K. Hamal, *Appl. Phys. Lett.* **12**, 148 (1968); K. Hamal, C. DeMichelis, and A. J. Alcock, *Opt. Acta* **16**, 463 (1969).
[26] J. H. Kiefer and R. W. Lutz, *Phys. Fluids* **8**, 1393 (1965).

the plasma image unaffected, and the various rays transmitted through the crossed slits seek out their correct image locations in the horizontal plane. Where these image locations correspond to places of horizontal deviation in the plasma there will be vertical displacements of the image due to the vertical imaging by the cylinder lens. A horizontal field mask over the plasma tube which selects a band of plasma is necessary to avoid confusion from the out-of-focus condensed vertical plasma image of the whole plasma which would otherwise be produced by the cylindrical lens. The result, which superficially resembles an oscilloscope trace, is a plot of horizontal refractive gradient vs horizontal position. Full-field schlieren and double-inclined-slit records are illustrated in Fig. 6 for two separate times in a θ-pinch discharge in nitrogen.

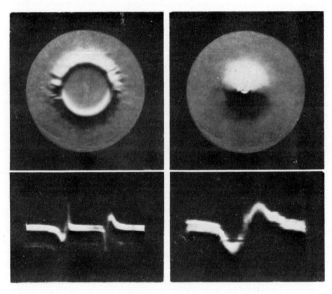

Fig. 6. Full-field schlieren and double-inclined-slit records for two times in nitrogen θ-pinch. (Courtesy of R. H. Lovberg, Univ. of California at San Diego, La Jolla, California.) [R. H. Lovberg, *AIAA J.* **4**, 1215 (1966).]

11.3.2.3.3. Shadowgrams. If, without the imaging lens L_2 (Fig. 4), a screen is inserted at a distance l past the refractive disturbance, the intensity distribution on the screen will differ in the presence of the disturbance from the otherwise uniform illumination. The luminous flux which would have illuminated an increment $\Delta x = x - x'$ is spread from $x + l\,\Delta\alpha$ to $x' + l\,\Delta\alpha$ for the case of constant gradient, or from $x + l\,\Delta\alpha$ to $x' + l(\Delta\alpha + \Delta x\,\partial\,\Delta\alpha/\partial x)$ when the gradient itself is changing. Thus, considering the similar relation in the y dimension, there is in the first case only a displacement but no change in area, whereas in the second case there is also a change in area, ΔA, which will result in an equivalent fractional change in illumination, ΔI,

$$\frac{\Delta I}{I} = \frac{\Delta A}{A} = l\left[\left(\frac{\partial \Delta\alpha}{\partial x} + \frac{\partial \Delta\alpha}{\partial y}\right)\bigg/\Delta x\, \Delta y\right]\Delta x\, \Delta y.$$

Combining the latter with Eq. (11.3.12) and again integrating over a sum of differentials gives

$$\frac{\Delta I}{I} = l\int_0^z \frac{1}{\mu}\left(\frac{\partial^2 \mu}{\partial x^2} + \frac{\partial^2 \mu}{\partial y^2}\right) dz. \tag{11.3.14}$$

The derivation of this result has assumed linearity in the change of the gradient in the transverse direction. As with all these optical methods, unfolding of the integrated result depends on a uniform or very simple z dependence. Furthermore, quantitative results are obtained only if the deviation of a ray from its undisturbed position on the screen is small enough to prevent confusion by overlapping of rays from various source locations.[27] This is generally not valid, so that shadow methods are primarily used to indicate the positions of rapidly changing density rather than for quantitative evaluation. The same considerations apply also in the "focused shadowgraph" methods in which the disturbance region is projected with a lens to a new image location while the observation plane remains at some nonzero distance from the image plane.

The great advantage of shadowgraphs, somewhat compensating for the difficulty of quantitative evaluation, is their experimental simplicity. Only the projection of a well-collimated beam on a detector possessing spatial resolution (photographic plate or film) is required. Two groups[28,29] have applied the method to plasmas with giant pulse lasers as the light source. Quantitative considerations of the required brightness and collimation[23] indicate that presently available conventional sources are inadequate.

If the density gradients are varying only slowly, producing minimum overlap of adjacent rays and weak intensity fluctuations, quantitative evaluations of the angular deflections in the shadowgraph arrangement may be made by observing the distortion in the shadow projection of a suitable grid placed in the collimated beam either before or after the region of refractive disturbance.[29] Simple geometric considerations

---

[27] D. W. Holder and R. J. North, "Schlieren Methods," Great Brit. Nat. Phys. Lab. Notes Appl. Sci. No. 31, p. 30. H.M. Stationery Office, London, 1963.

[28] U. Ascoli-Bartoli, S. Martellucci, and E. Mazzucato, *Proc. Int. Conf. Ioniz. Phenomena Gases, 6th*, 1963, **4**, p. 105. S.E.R.M.A., Paris, 1964.

[29] F. C. Jahoda, E. M. Little, W. E. Quinn, F. L. Ribe, and G. A. Sawyer, *J. Appl. Phys.* **35**, 2351 (1964).

relate the distortion to the net angular deflection. The sensitivity of the grid projection technique is enhanced by observation of the distortions in the moiré fringes produced when two rather fine scale grids at some suitable separation and orientation are inserted in the collimated beam.[30,31] An angular deviation in any part of the field sufficient to displace the shadow projection of a particular ruling on the first plate by one period of the rulings on the second plate results in a local shift of the moiré pattern by one fringe.

A new quantitative method that uses the spherical abberation of a concave mirror to combine features of the double-inclined-slit schlieren method with shadowgraphs to give a full view of the test section has been reported.[32]

11.3.2.4. Sensitivity Limits. The ultimate limitation on all these methods after the more practical problems such as adequate light sources, high quality lenses and windows, and critical alignment have been met, is due to diffraction. While the concepts have all been described in terms of geometrical optics, diffraction effects can become large because the pertinent aperture limits are not the lens apertures but rather the sizes of the refractive disturbances themselves. These limits are comparable[21,22,33] to those of interferometry with due regard to the fact that schlieren methods measure density gradients while interferometry measures density and the particular plasma distribution may therefore favor one or the other.

The direct utilization of diffraction phenomena in the phase contrast methods pioneered by Zernike is closely related to schlieren methods and might be treated under the same heading, as has been done by Wolter[20] in one of the important reviews of the subject. Since, however, phase contrast methods have not yet been widely applied to plasmas we do not extend our discussion that far, except to mention that one group[34,35] has indeed studied plasma phase objects claimed to be invisible to schlieren photography.

11.3.3. Faraday Rotation

11.3.3.1. Magnitude of Rotation. In Section 11.2.1.3 it was shown that in the presence of a static axial magnetic field there are two distinct propaga-

[30] Y. Nishijima and G. Oster, *J. Opt. Soc. Amer.* **54**, 1 (1964).
[31] B. J. Rye, J. W. Waller, A. S. V. McKenzie, and J. Irving, *Brit. J. Appl. Phys.* **16**, 1404 (1965).
[32] S. Knöös, *J. Plasma Phys.* **2**, 243 (1968).
[33] P. B. Barber, D. A. Swift, and B. A. Tozer, *Brit. J. Appl. Phys.* **14**, 207 (1963).
[34] H. M. Presby and D. Finkelstein, *Rev. Sci. Instrum.* **38**, 1563 (1967).
[35] H. M. Presby and R. J. Meehan, *Phys. Fluids* **11**, 1487 (1968).

tion velocities through a plasma associated respectively with the two **E**-field amplitudes that were defined by $E_\pm = E_x \pm iE_y$. By these definitions,

$$E_x = \tfrac{1}{2}(E_+ + E_-)$$
$$E_y = (1/2i)(E_+ - E_-) = -(i/2)(E_+ - E_-). \tag{11.3.15}$$

Writing out explicitly the oscillatory exponential terms, we obtain

$$E_x = \tfrac{1}{2}E_+ \exp i(k_+ z - \omega t) + \tfrac{1}{2}E_- \exp i(k_- z - \omega t)$$

and

$$E_y = -(i/2)E_+ \exp i(k_+ z - \omega t) + (i/2)E_- \exp i(k_- z - \omega t), \tag{11.3.16a}$$

whose real parts are

$$E_x = (E_+/2)\cos(k_+ z - \omega t) + (E_-/2)\cos(k_- z - \omega t),$$
$$E_y = (E_+/2)\sin(k_+ z - \omega t) - (E_-/2)\sin(k_- z - \omega t). \tag{11.3.16b}$$

Consider an electric wave of unit amplitude, linearly polarized in the x direction, incident at the magnetized plasma boundary $z = 0$. These boundary conditions require $E_+ = E_- = 1$. By trigonometric identities, the x and y components at arbitrary distance z into the plasma are

$$E_x = \cos[\tfrac{1}{2}(k_+ + k_-)z - \omega t]\cos[\tfrac{1}{2}(k_+ - k_-)z],$$
$$E_y = \cos[\tfrac{1}{2}(k_+ + k_-)z - \omega t]\sin[\tfrac{1}{2}(k_+ - k_-)z]. \tag{11.3.16c}$$

The two components therefore oscillate in phase (linear polarization) but the last factor in each term indicates that as z increases the maximum amplitudes of the x and y components vary sinusoidally with a 90° phase difference. This means that the plane of the linear polarization at $z = l$ is rotated by

$$\theta = \tfrac{1}{2}(k_+ - k_-)l \text{ rad}. \tag{11.3.17}$$

Since a right-hand coordinate system and **B** along $+z$ were assumed in the equations of motion (11.2.20), and it follows from Eq. (11.2.26) that $k_+ > k_-$, the rotation is counterclockwise looking toward the source of radiation at negative z and is in the same sense as electrons gyrate about the magnetic field.

It is instructive to digress for a simple physical interpretation of the sense of rotation. The linear polarization is decomposed into two oppositely rotating circular polarizations. In the magnetized medium one of these rotates in the same sense that electrons gyrate. This must be the component that shows a resonance effect when the propagation frequency and the gyrofrequency become equal, and is therefore identified with the

k_- portion of (11.2.26). Since $k_- < k_+$, its associated phase propagation velocity, $c_- = \omega/k_-$, is greater than c_+. At a fixed position in space, the faster wave is leading in phase relative to the slower wave, and the net rotation of the linear polarization will also be in the sense of rotation of the faster wave (rotation with the electrons).

If the light were reflected back through the magnetized plasma, the sign of B would be reversed and the rotation would remain counterclockwise if the observer did not change position, i.e., would view along rather than against the direction of propagation. Thus the net rotations of multiple traversals are additive.

Using only the first-order terms in the power expansion of (11.2.26) in the small parameters ω_p/ω and ω_e/ω to evaluate $k_+ - k_-$, Eq. (11.3.17) becomes

$$\theta = \omega_e \omega_p^2 l / 2c\omega^2$$
$$= (1/2\pi)(e^3/m_e^2 c^4)\lambda^2 n_e Bl$$
$$= 2.63 \times 10^{-17} \lambda^2 n_e Bl \text{ rad}, \qquad (11.3.18)$$

where the numerical value of the constant is appropriate when the wavelength λ and the path length l are given in centimeters, the unit of electron density n_e is cm^{-3} and the axial magnetic field B is in gauss.

The substitution of numerical values in (11.3.18) shows that the rotation is only a small effect at optical wavelengths even for substantial densities and fields. For instance, at the He–Ne laser wavelength of 6328 Å, for a 10-kG field, 1-meter path length and 10^{17} electrons/cm^3 the rotation is 36' of arc. Because of the λ^2 dependence it would seem a great advantage to use the larger rotations resulting from using infrared laser wavelengths. However, density gradients perpendicular to the beam direction cause beam deflections that, as seen in Eq. (11.3.13a), also scale with λ^2. The advantage of increased rotation angle may in practice be more than offset by the increased refractive beam deviations, particularly in view of the small size of infrared detectors and the large variation of spatial sensitivity over their surface.

11.3.3.2. *Measurement of Rotation.* Rotations of the direction of linear polarization are measured by observing the intensity variation transmitted by a polarization analyzer of fixed orientation. The transmitted intensity is proportional to $\cos^2 \alpha$, where α is the angle between the polarization to be measured and the maximum pass direction of the analyzer. If a transient plasma modulates a steady laser source, thereby permitting measurement of the ac component change in the signal level, the greatest sensitivity is achieved at $\alpha \sim 45°$. If, on the other hand, a transient laser pulse illuminates a plasma that remains constant during the light pulse duration, then the total transmitted laser signal is measured and the greatest fractional change (comparing the case of plasma present vs no plasma) will occur for minimum total transmission, i.e., near $\alpha = 90°$. In either case the sensitivity can be doubled by using two analyzers set at positive and negative rotations to the unrotated polarization, respectively, and subtracting the two results. Subtraction has the further advantage of canceling intensity variations not due to rotation of the plane of polarization. The sum of the two signals can serve as the normalization needed to determine the rotation angle θ from the difference signal. The angle θ will actually give an averaged value over the total path length $\theta \sim \int_0^l B(z)n_e(z)\,dz$ from which either average B or average n can be deduced if independent knowledge of the other parameter is available.

Faraday rotation measurements of plasmas have been reported for continuous wave He–Ne lasers operating at 6328 Å[36,37] and at 3.39 μ,[38,39] and for two types of pulsed lasers, the ruby laser at 6943 Å[40] and the water vapor laser at 28 μ.[41]

The greatest sensitivity has been achieved at 6328 Å, where rotations as small as 0.01° have been observed. Gribble et al.[37] achieved this with large area, uniform sensitivity photoconductive silicon detectors. A schematic diagram of their experimental arrangement, which serves to illustrate principles fundamental to the other experiments also, is shown in Fig. 7. The Wollaston prism spatially separates two mutually orthogonal linear polarizations, and is oriented so as to make the two signals approximately equal. More precise equality in the absence of plasma is achieved by individual gain adjustment of the signal amplifiers. Lens A focuses the laser beam at the plasma midplane while lens B reduces divergence due to refraction and spreads the signals over relatively large areas on the detectors. The polarization rotator is used to check that the difference signal reverses polarity for a 90° rotation of the incident plane of polarization. The purpose of this experiment was to measure magnetic field strength, and the results were normalized to electron density measurements made independently with an interferometer. The minimum sensitivity was $6/f$ kG, where f is the number of interference fringes at the ruby laser wavelength. Falconer and Ramsden[36] report a similar sensitivity and give a

[36] I. S. Falconer and S. A. Ramsden, *J. Appl. Phys.* **39,** 3449 (1968).

[37] R. F. Gribble, E. M. Little, R. L. Morse, and W. E. Quinn, *Phys. Fluids* **11,** 1221 (1968).

[38] A. A. Dougal, J. P. Craig, and R. F. Gribble, *Phys. Rev. Lett.* **13,** 156 (1964).

[39] P. Bogen and D. Rusbüldt, *Phys. Fluids* **9,** 2296 (1966).

[40] P. H. Grassmann and H. Wulff, *Proc. Int. Conf. Ioniz. Phenomena Gases*, 6th, 1963, **4,** p. 113. S.E.R.M.A., Paris, 1964.

[41] A. N. Dellis, W. H. F. Earl, A. Malein, and S. Ward, *Nature (London)* **207,** 56 (1965).

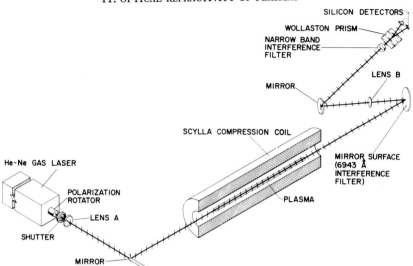

FIG. 7. Diagram of Faraday rotation apparatus. (Courtesy of R. F. Gribble, Los Alamos Sci. Lab.) [R. F. Gribble, E. M. Little, R. L. Morse, and W. E. Quinn, *Phys. Fluids* **11**, 1221 (1968).]

detailed discussion of optical systems, detectors, and beam deviation problems.

11.4. Experimental Methods Requiring Laser Sources

11.4.1. Gas Laser Interferometry

11.4.1.1. Coupled Cavity Interferometry. Coupled cavity interferometry is a new technique, based on the discovery[42] that if a small fraction of a gas laser's output radiation is returned into the laser with an external mirror, the intensity of the laser output is modulated. If the light fed back is in phase with the laser, the laser oscillation amplitude is reinforced; if the light fed back is out of phase, the laser amplitude is depressed. These amplitude or intensity modulations are produced by laser gain changes as will be explained below.

A schematic diagram of the gas laser interferometer is shown in Fig. 8. The output beam of the gas laser is directed through the plasma to be measured and is returned by mirror M_3 back through the plasma into the laser. The amplitude modulation can be explained as an effective variable reflectivity of the laser mirror M_2. Both mirrors M_2 and M_3 return light into the laser cavity. When the light from M_3 is in phase with the light

---

[42] P. G. R. King and G. J. Steward, *New Sci.* **17**, 180 (1963).

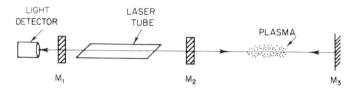

FIG. 8. Diagram of gas laser interferometer.

from M_2, its effect is equivalent to increasing the reflectivity of mirror M_2. The increase in light fed back into the cavity reduces the loss term in the laser gain equation and the output intensity increases. When the light from M_3 is returned out of phase, the effective reflectivity is reduced and the laser intensity is depressed. Due to the double pass of the light through the cavity, successive complete cycles of intensity variation (fringes) are produced as the length of the external cavity changes by $\frac{1}{2}$-wavelength increments. The fringes are unsymmetrical, however, because the gain of the laser is a nonlinear function of the mirror reflectivity. When a plasma of length l is introduced into the cavity, the refractivity of the electrons will produce an effective path length change equivalent to a number of fringes

$$N = (\mu - 1)2l/\lambda. \tag{11.4.1}$$

11.4.1.2. Sensitivity of the Coupled Cavity Interferometer. From Eqs. (11.2.16) and (11.4.1)

$$N = (\omega_p/\omega)^2 l/\lambda$$
$$= 8.98 \times 10^{-14} n_e l \lambda \text{ in cgs units.}$$

The number of fringes produced is directly proportional to the wavelength used. At the He–Ne gas laser infrared wavelength $\lambda = 3.39\ \mu$ the sensitivity of the interferometer is

$$N = n_e l/3.3 \times 10^{16}. \tag{11.4.2}$$

Fringe shifts can be measured to only about $\frac{1}{4}$ fringe, because of the unsymmetrical character of the fringes. This coupled cavity interferometer is useful, therefore, for plasmas whose integrated densities in the line of sight are in the range 10^{16}–10^{18} electrons/cm^2. However, extension of the technique to lower density is described in Section 11.4.1.5.

11.4.1.3. Application to Pulsed Plasmas. The technique is directly suitable only for pulsed plasmas where time variation of the plasma density can be followed along a single line of sight. The spatial resolution is equal to the laser beam width. As the plasma is generated, its increasing electron

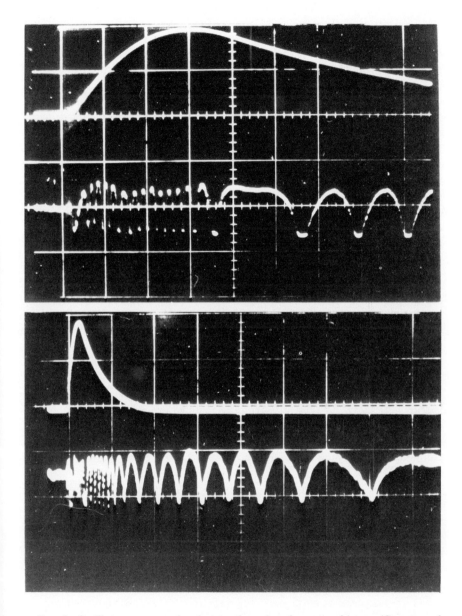

FIG. 9. Oscilloscope traces showing gas laser interferometer fringes. (Courtesy of D. E. T. F. Ashby, Culham Laboratory, England.)

density is equivalent to a decreasing cavity length which modulates the laser output. The modulated laser output is detected by a photomultiplier or other photo detector, appropriate for the laser wavelength used. Figure 9 is an oscilloscope trace showing a typical modulated laser output. This example is from the original work of Ashby and Jephcott who first applied the method to plasma diagnostics.[43] The laser was operating at 3.39 μ and 0.6328 μ simultaneously. A germanium filter allowed only 3.39-μ radiation into the plasma cavity so that the interference effects were produced at 3.39 μ. However, the detection was done at 0.6328 μ with a photomultiplier. Since the two transitions are coupled through a common upper level, their outputs have a reciprocal relationship. It is not possible to distinguish increasing plasma density from decreasing plasma density in this system, except by noting the time symmetry. In Fig. 9, the plasma density starts from zero, rises to a maximum near the center of the trace and then decreases to zero again at the end of the trace. The sense reversal occurs during the longest period.

11.4.1.4. Frequency Response. In more recent work, both interference and detection are done at 3.39 μ, because higher frequency response can be achieved.[44,45] Because the laser gain for the 3.39-μ line is much greater the laser mirror reflectivities for this wavelength are much smaller than for 6328 Å in order to prevent complete domination by the infrared line. This has the important effect of lowering the laser Q at 3.39 μ, enabling faster response, and increasing the coupling to the external plasma. Typically, it is found that the upper frequency limit for the 6328-Å transition is 1 or 2 MHz whereas at 3.39 μ, the upper frequency limit is above 20 MHz. At very high fringing rates where the change in external cavity length approaches a wavelength during one double pass of the light, spurious fringes may be introduced.[46] Gas laser interferometers of the coupled cavity type have been reported at the three He–Ne wavelengths, 0.6328μ,[34] 1.15 μ,[47] and 3.39 μ,[43,44] and at 10.6 μ in a CO_2 laser.[48]

When it is desired to use the infrared 3.39-μ transition, a conventional He–Ne laser with mirrors coated for the 6328-Å red transition can be used, provided only that the mirrors are coated on a quartz substrate that will transmit the infrared light. When the external cavity mirror, usually a metallic mirror that reflects both red and infrared light, is

---

[43] D. E. T. F. Ashby and D. F. Jephcott, *Appl. Phys. Lett.* **3**, 13 (1963).
[44] D. A. Baker, J. E. Hammel, and F. C. Jahoda, *Rev. Sci. Instrum.* **36**, 395 (1965).
[45] R. F. Gribble, J. P. Craig, and A. A. Dougal, *Appl. Phys. Lett.* **5**, 60 (1964).
[46] J. H. Williamson and S. S. Medley, *Can. J. Phys.* **47**, 515 (1969).
[47] J. B. Gerardo and J. T. Verdeyen, *Proc. IEEE* **52**, 690 (1964).
[48] H. Herold and F. C. Jahoda, *Rev. Sci. Instrum.* **40**, 145 (1969).

brought into alignment, the gain in the infrared is raised enough so that the infrared transition dominates and the red light generally extinguishes completely. The red light is very useful in locating the infrared beam path during alignment, and its disappearance, in fact, is a convenient indicator of precise alignment.

11.4.1.5. Extension of Sensitivity Limit. The lower limit of density sensitivity of gas laser interferometry has been considerably extended beyond that given in Section 11.4.1.2 by modifications of the basic geometry shown in Fig. 8. Three different methods are useful.

11.4.1.5.1. ROTATING FEEDBACK MIRROR. If mirror M_3 is put into uniform motion toward or away from the laser, it superimposes a uniform modulation on the plasma effect which increases the sensitivity to density variations about a factor ten.[44] Figure 10 shows a system of mirror translation using a rotating roof mirror and a stationary mirror. This

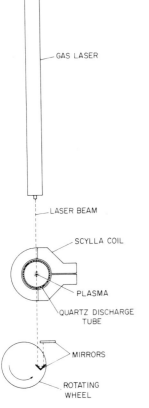

FIG. 10. Gas laser interferometer apparatus. The rotating mirror introduces continuously varying phase in the feedback path.

arrangement allows the light to be returned accurately into the laser over a 30° rotation angle of the mirror. The mirror alone produces a nearly constant frequency modulation of the laser. When the plasma is added, the frequency of the moving mirror modulation is altered slightly.

Measurement of successive periods of the oscillation, with and without plasma, allows determination of fringe shifts as small as 1/20 fringe, extending the lower limit of sensitivity to 10^{15} electrons/cm² at $\lambda = 3.39\ \mu$.

An additional advantage of the rotating mirror system is that it provides positive indication of the sense of the plasma density change, according to whether the period of the oscillation is increased or decreased by the plasma effect for a given moving mirror direction.

11.4.1.5.2. OSCILLATING MIRROR. A further increase in the sensitivity can be made by attaching mirror M_3 to a speaker coil and oscillating it sinusoidally at about 1 kHz through a known amplitude of 0.1 λ or less.[49] When this is done, plasma effects small compared to the imposed motion can be measured.

The amplitude of oscillation of the speaker coil must be calibrated in advance. This is usually done by using multiple reflections in the interferometer to obtain at least one fringe displacement, so that the motion can be calibrated in terms of full fringes. Having made this calibration, the oscillating mirror is used as the feedback mirror in a plasma measurement. The modulation effect of the known mirror oscillation amplitude will vary as random drifts of the external cavity length change the phase of the returning light. Therefore, the laser modulation amplitude, due to the known mirror oscillation, must be determined each time simultaneously with the plasma measurement. The laser signal is displayed on two oscilloscope beams. One beam is set to display the imposed modulation of the oscillating mirror. The plasma effect is barely visible on the beam as a small wiggle on this trace. The other beam is set at a known higher gain and faster sweep speed to enlarge the plasma effect superimposed on the speaker oscillation. Changes in cavity length produced by the plasma appear as amplitude variations on the laser output, which by comparison with the calibration amplitude can be interpreted as known fractions of a fringe. Fringe shifts as small as 10^{-3} fringe have been measured with a CO_2 laser when the mirror was oscillated through 0.01 wavelength.[48] In this case, with the additional gain in sensitivity due to the 10-μ CO_2-laser wavelength, 10^{-3} fringe corresponds to an integrated density of 1.1×10^{13} electrons/cm².

In order to apply this method it is important that the laser amplitude response be the same, for a given cavity length change, at both the calibra-

---

[49] K. S. Thomas, *Phys. Fluids* **11**, 1125 (1968).

tion signal frequency and the plasma signal frequency, or that at least a correction factor is determined by an amplitude versus frequency calibration.

Mechanical vibration of the mirrors will produce spurious phase changes and false density indications which are often a severe problem in sensitive systems. One solution is to operate only at relatively vibration-free instants of time. An ingenious method[50] used with a 3.39-μ coupled cavity interferometer involves use of a second laser operating at 0.6328 μ to monitor vibration so that the vibration effect can be subtracted out.

11.4.1.5.3. OFF-AXIS MODES. Increase in sensitivity of a gas laser interferometer has been reported when transverse laser modes are used in addition to axial modes.[51] This requires that mirror M_3 be spherical. The transverse modes can be regarded as equivalent to off-axis rays which make multiple traversals of the plasma cavity before reentering the laser.

11.4.1.6. Heterodyne Interferometry

11.4.1.6.1. PRINCIPLE OF OPERATION. A different method of gas laser interferometry is achieved by placing the plasma to be measured inside the laser cavity instead of in an external cavity.[52] Changes in the cavity length produced by the plasma alter the frequency of the laser oscillation in this arrangement, instead of modulating the amplitude as in the previous arrangement. The changes in laser frequency are detected by mixing the signal with the signal from a second local-oscillator laser, which operates at fixed frequency, and observing the beat frequency.

11.4.1.6.2. SENSITIVITY. The heterodyne method is applicable to plasmas of low density, 10^{10}–10^{13} electrons/cm^2, as can be seen from the following analysis of the sensitivity. The resonance condition for the axial modes of the laser without plasma ($\mu = 1$) is given by

$$N\lambda_1/2 = l, \tag{11.4.3}$$

where l is the length of the cavity and N the axial mode number. Single mode operation is assumed. If a plasma of length d is inserted into the laser cavity, the resonant condition becomes

$$N\lambda_2/2 = l + (\mu_p - 1)d = l - \tfrac{1}{2}[\omega_p^2/\omega^2]d. \tag{11.4.4}$$

[50] A. Gibson and G. W. Reid, *Appl. Phys. Lett.* **5**, 195 (1964).
[51] J. T. Verdeyen and J. B. Gerardo, *Appl. Opt.* **7**, 1467 (1968).
[52] W. B. Johnson, A. B. Larsen, and T. P. Sosnowski, *Proc. Int. Conf. Ioniz. Phenomena Gases, 7th, 1965*, **3**, p. 220. Gradevinska Knija Publ., Belgrade, 1966.

When Eq. (11.4.4) is subtracted from Eq. (11.4.3) corresponding to heterodyning the two lasers,

$$(N/2)\,\Delta\lambda = \tfrac{1}{2}(\omega_p^2/\omega^2)\,d \tag{11.4.5}$$

or, converting to change in frequency $v = \omega/2\pi$ and using (11.4.3),

$$\Delta v = -\frac{v}{\lambda}\Delta\lambda = \frac{1}{4\pi}\frac{\omega_p^2}{\omega}\frac{d}{l} = \frac{n_e e^2 d}{m_e \omega l}$$

$$= \frac{n_e e^2 \lambda d}{2\pi m_e c l}. \tag{11.4.6}$$

The frequency shift is directly proportional to the electron density and to the laser wavelength. At the red He–Ne wavelength $\lambda = 0.6328\,\mu$, a plasma with density $n_e = 10^{13}$ electrons/cm^3 and of such a length that it fills $\tfrac{1}{10}$ of the laser cavity, will shift the laser frequency 86 kHz. This is a convenient beat frequency to detect. This method has higher sensitivity than the regular gas laser interferometer. However, the plasma effect to be measured must be of long enough duration so that the beat frequency can be observed. For instance, in the example above, an 86-kHz beat would require several cycles or several tens of microseconds duration to be observed. Higher sensitivities, corresponding to lower beat frequencies, would require correspondingly longer observation times. This requirement implies that only rather slowly varying, stable plasmas can be measured by this method.

11.4.1.6.3. MECHANICAL STABILITY. The success of this very sensitive measuring technique depends on very good frequency stability of the two lasers. The frequency shift to be measured in the example above is about one part in 10^{10} of the laser frequency. Or to put it in terms of mechanical motion, a 1.7-Å change in the length of the laser cavity, will produce a shift of 86 kHz, the same as the shift calculated in the previous paragraph. In practice, mechanical vibration has been a serious problem in heterodyne interferometry. Antivibration mounting of the entire system, and quiet environments, have been absolutely necessary.

11.4.1.7. Gas Laser Interferometry without Coupling of Laser and Plasma

11.4.1.7.1. FABRY–PEROT INTERFEROMETRY. It is possible to place the plasma to be measured between the mirrors of a Fabry–Perot etalon and make a sensitive measurement of plasma density through changes in the Fabry–Perot transmission. Korobkin and Malyutin[53] have successfully

---

[53] V. V. Korobkin and A. A. Malyutin, *Sov. Phys. Tech. Phys.* **13**, 908 (1969).

applied this technique to measure densities $\sim 10^{14}/\mathrm{cm}^3$. They used a piezoelectric transducer and feedback of the transmitted light to stabilize the Fabry–Perot cavity against long term drifts. A $\frac{1}{4}$-wave plate was placed between the laser and the Fabry–Perot cavity to rotate the plane of polarization of the light and prevent phase-shifted light fed back into the laser from affecting its output.

11.4.1.7.2. WAVELENGTH EXTENSION OF MACH–ZEHNDER INTERFEROMETRY. Three groups[54-57] have reported using an HCN gas laser source at 337 μ in a Mach–Zehnder configuration for plasma density measurements. The plasma is probed along the single line of sight set by the laser beam diameter with fast time response using Golay cells or liquid-He cooled InSb detectors. The sensitivity for a single pass through the plasma becomes $6.6 \times 10^{14}/\mathrm{cm}^2$ per fringe. It seems likely that the coupled cavity techniques would work at these wavelengths also. Parkinson et al.[56] have used the laser attenuation observed while measuring the phase shift to deduce a collision frequency and thus electron temperature (Section 11.2.2).

11.4.2. Holographic Phase Measurements

11.4.2.1. Introduction to Holography.[58,59] A hologram stores both the amplitude and phase of an optical field incident on the hologram plane in a manner that permits a faithful recreation at later times of the original phase and amplitude distribution. The reconstructed wavefront may then be used instead of the original to determine, for instance, the spatial phase variations.

Holography first became practical with the availability of the highly coherent radiation emitted by laser sources. A hologram exposure is an interferogram between (1) a "scene" beam reaching the hologram plane (photographic plate) by reflection from or, as in the case of plasma, transmission through the object, and (2) a reference beam, originating in the same source and reaching the hologram plane by a path that circumvents the object (Fig. 11).

---

[54] S. Kon, M. Otsuka, M. Yamanaka, and H. Yoshinaga, *Jap. J. Appl. Phys.* **7**, 434 (1968).
[55] R. Turner and T. O. Poehler, *J. Appl. Phys.* **39**, 5726 (1968).
[56] G. J. Parkinson, A. E. Dangor, and J. Chamberlain, *Appl. Phys. Lett.* **13**, 233 (1968).
[57] J. Chamberlain, H. A. Gebbie, A. George, and J. D. E. Beynon, *J. Plasma Phys.* **3**, 75 (1969).
[58] E. N. Leith and J. Upatnieks, *J. Opt. Soc. Amer.* **54**, 1295 (1964).
[59] H. M. Smith, "Principles of Holography." Wiley (Interscience), New York, 1969.

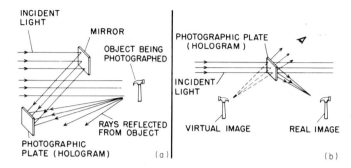

FIG. 11. (a) Diagram of hologram recording and (b) hologram reconstruction.

Expressed analytically the intensity at the plate is

$$I = (E_r e^{i\alpha x} + E_s e^{i\varphi(x)})(E_r e^{-i\alpha x} + E_s e^{-i\varphi(x)})$$
$$= E_r^2 + E_s^2 + E_r E_s e^{i\varphi(x) - i\alpha x} + E_r E_s e^{-i\varphi(x) + i\alpha x}. \quad (11.4.7)$$

E_r is the amplitude of the reference beam, E_s is the amplitude of the scene beam, both assumed real, and x is the coordinate in the hologram plane (taken one-dimensionally for notational convenience only). The amplitudes E_s and E_r of both beams are assumed constant, while the phase φ of the scene beam is an unspecified function of position, and the phase of the reference beam, assumed a plane wave incident at angle $\theta = \sin^{-1}(\lambda/2\pi)\alpha$, is linearly proportional to the hologram coordinate. The time factors have been omitted in (11.4.7) because oscillations at optical frequencies are never resolved and all our considerations apply to the envelopes of amplitude or intensity. For incoherent beams the cross-product terms would average to zero over any realizable observation time due to additional randomly varying relative phases between the two beams. The greater the temporal and spatial coherence of the laser source the less restrictive, respectively, are the requirements on path equality of the two beams and on accurate point-for-point overlay of the two wavefronts at the photographic plate. These are just the same requirements as those encountered in Mach–Zehnder interferometry. However, because the mean angle between the two beams will generally be several degrees, the fringe structure is exceedingly fine. In consequence, the emulsion recording the hologram must be capable of high resolution. (It is important to distinguish this fine scale intensity modulation on the photographic plate, produced by conventional interference in the construction of any one hologram, from the gross interference pattern due to the beating together of two similar holograms produced by two consecutive exposures that is the subject matter of holographic interferometry as discussed in Section 11.4.2.2.)

The recorded photographic density is linearly proportional to the intensity I for variations small compared to the constant background term E_r^2 ($E_r \gg E_s$) and the amplitude transmittance of the developed emulsion becomes $T(x) = T_0 - kI(x)$. When the hologram is reconstructed by illuminating it with a plane wave reconstruction beam (Fig. 11b) the transmitted amplitude contains a term, among others, proportional to the third term in (11.4.7). This term itself is proportional to the original subject waveform deflected by the linear phase term $e^{-i\alpha x}$ (analogous to ordinary grating deflection into 1st order) in the direction $-\theta$. Thus an observer looking through the hologram toward the reconstruction beam, as through a window, will see a virtual image of the original scene angularly separated from the reconstruction beam direction. There is another term in the positive θ direction which duplicates the original subject phase distribution except for reversal in sign. This is equivalent to the original wavefront traveling with reversed sign in time and thereby converging to a real image in front of the plate. Since the absolute sign of the phase is of no concern, either the real or virtual reconstructions can be used according to convenience.

In practice a great deal of latitude in relative beam intensities and processing procedure is permissible. This is particularly so in the interferometric application where only the relative differences between two exposures are of primary interest. If the hologram exposure falls outside the range of linear film response, the nonlinearities primarily generate higher order terms that are deviated into higher multiples of the angle θ, forming additional higher order images. The use of other than plane parallel reference and reconstruction beams or a change of wavelength between hologram formation and reconstruction mainly changes the image locations and magnifications.

The most restrictive requirements are on source coherence and detector resolution. For the former, even giant pulse ruby lasers without special mode selection are adequate if the reference and scene beams are kept equal in length to within a few centimeters and if a small external pinhole is used to improve the spatial coherence.[60] The pinhole, however, decreases the energy density reaching the film plane sufficiently (especially after the beam is expanded to a useful working aperture) to make it desirable to use a faster emulsion than the extremely slow, ultra-high resolution Kodak 649 emulsion generally used for CW holography. The Kodak SO243 emulsion is about 1000 times faster at the ruby wavelength, and its 500 line/mm resolution capability is adequate as long as the two beam directions do not have an angular separation greater than 15°. Intermediate

[60] F. C. Jahoda, R. A. Jeffries, and G. A. Sawyer, *Appl. Opt.* **6,** 1407 (1967).

between these in speed and almost comparable to Kodak 649 in resolution are the Agfa-Gevaert 8E75 and 10E75 plates.

11.4.2.2. Plasma Applications

11.4.2.2.1. INTERFEROMETRY.[61] If two sequential hologram exposures are made on the same photographic emulsion, the wavefront amplitude and phase structure of the scene beam from both exposures are recorded. When the double exposure hologram is reconstructed, the wave fields of the two exposures are reproduced simultaneously. The reconstructions are coherent since they originate in a common source and the two images will interfere with each other according to the phase differences between the two exposures. The result is analogous to a conventional interferogram. However, the wavefronts being compared originally existed at different times rather than, as in a conventional interferometer, existing simultaneously and requiring first a spatial separation and subsequent recombination. Since these wavefronts traverse a common path (aside from the difference purposely introduced between exposures, e.g., plasma) the experimental requirements are greatly simplified. In holographic interferometry there do not exist the needs of conventional interferometry for (a) precise alignment to achieve recombination of the beams and (b) for extremely high quality optical flats to prevent phase differences between the beams other than those one desires to measure in the plasma. Any kind of windows, in particular transparent walls of arbitrary curvature in the plasma vessel, are sufficient. The complexity associated with a conventional interferometer capable of spanning the plasma discharge is entirely dispensed with.

Sometimes substantial advantage may be realized by inserting a diffuser into the scene beam between the plasma and the hologram plane. The advantage of the diffuser is that each point of its surface becomes a secondary source which illuminates all points of the hologram plane. In consequence, reconstruction of any small portion of the hologram will reproduce the entire scene as projected onto the diffuser with only eventual loss of resolution in the reconstruction as the hologram area decreases. Equivalently, even if both beams overlap over only a small area of the hologram, the complete information is encoded there.

The diffuser may also be placed in the scene beam before it traverses the plasma instead of after. The plasma is then illuminated over a whole range of angles and the double exposure holographic recording will give changing fringe patterns (and thus integrated density distributions)

[61] L. O. Heflinger, R. F. Wuerker, and R. E. Brooks, *J. Appl. Phys.* **37**, 642 (1966).

corresponding to different viewing directions in the play-back. If the geometry permits viewing at sufficiently different angles this allows some inference about the depth dimension. In this case it is essential to restrict the viewing angle for any given direction in the reconstruction to prevent complete smearing out of fringes due to overlapping views.

An advantage of operating without the diffuser is that the fringes are not localized in space and will appear on the hologram itself when it is viewed at glancing angle in incoherent nonmonochromatic light. (It is instructive in this case to recognize that the dark fringes are just the moiré beats produced where the fine scale grating of one exposure is out of registry with the grating structure of the other exposure.) By avoiding the speckle effect produced by reconstruction in laser light, better spatial resolution is achieved in cases where fringes crowd together. Also by overlapping corresponding portions of the reference and scene beams on the hologram, the spatial coherence requirement on the laser source is further reduced.

The double exposure holographic interferogram described so far is actually the analog of the conventional interferometer adjusted for a uniform field, the case sometimes called the infinite fringe spacing. This adjustment is not generally desirable because it is not possible to detect phase disturbances small compared to a full fringe. Usually, the conventional interferometer is adjusted for a straight line fringe background by introducing a slight angle between the recombined beams with a small tilt of at least one mirror or beam splitter. In the holographic case the background fringes are easily produced by purposely making a linear phase change between exposures. This is readily done either with a hollow wedge inserted in the optical path of the scene beam whose filling gas is changed between exposures or by a mechanical motion of one mirror between exposures.

In practice this deliberately introduced background serves another extremely useful function.[60] If one does not have good antivibration mounting of the entire holographic optical system, spurious phase shifts are introduced by relative motions between various components that have occurred during the time between the two exposures. In general the fringes due to vibration are linear over the field of view. If the wedge is designed to produce a large number of background fringes compared to the number of fringes produced by the random vibration, the vibration produced fringes alter only slightly the orientation and spacing of the background fringes. Without the controlled background the random motions would produce a background of wholly random spacing and orientation. If this spacing is not dense enough, it might not even be recognized as spurious and the extra phase changes would then be mistakenly added to the phase changes attributed to the plasma.

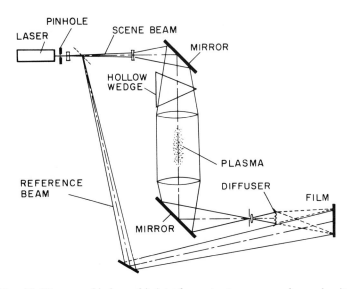

FIG. 12. Diagram of holographic interferometer to measure plasma density.

Figure 12 is a schematic of the experimental arrangement used in the application of this technique to a plasma generated by a plasma gun located in the center of a 1.2-meter diameter vacuum tank.[62] The tank could not readily have been spanned by a conventional interferometer. The scene beam is expanded to cover 25-cm windows (also an extreme dimension for conventional interferometry), and after passing through the plasma and the hollow wedge, is again focused down to pass through an aperture which prevents most of the plasma self-luminosity from getting through. An auxiliary lens which images the plasma location on the diffuser counteracts the refractive bending effects of the plasma which otherwise would degrade the fringes in the region of the plasma boundary. The reference beam is made of "equal" length to the scene beam as measured with a meter stick. Two ruby giant pulse laser exposures are made separated in time by a minute or two, the first without plasma present and air in the wedge, the second synchronized with a plasma discharge and helium in the wedge. Typical results obtained with a processed doubly exposed hologram of a neon plasma leaving the muzzle of a coaxial gun at various times after gas breakdown are shown in Fig. 13. The pictures were recorded on film placed at the real image location during the reconstruction of the doubly exposed holograms with CW-laser light. Each

[62] T. D. Butler, I. Henins, F. C. Jahoda, J. Marshall, and R. L. Morse, *Phys. Fluids* **12**, 1904 (1969).

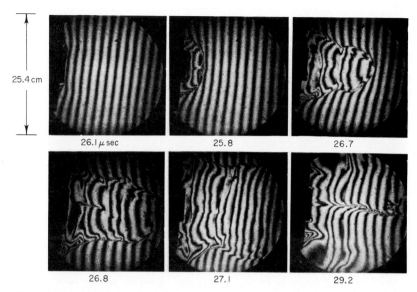

FIG. 13. Holographic interferograms of coaxial gun neon plasma at different times after gas breakdown.

frame corresponds to a different discharge, and the slightly different slants of the background fringes illustrate the result of slight uncontrolled motions discussed above.

Komisarova et al.[63] and Jeffries[64] have used both the ruby laser fundamental wavelength and its second harmonic produced in a KDP crystal simultaneously in holographic interferograms of a laser spark and exploding wire, respectively. The latter was able to separate quantitatively the electron refractivity from the density compression due to neutral particle pileup at the shock front. The common scene and reference beams each contain both colors, producing superimposed interferograms in the two colors. When the picture is reconstructed in monochromatic light, angular separation of the two interferograms is obtained because the average grating spacings produced by the two reference beam colors at a common angle of incidence are different.

As discussed above, for single frame interferograms holographic interferometry is experimentally simpler to implement than the conventional techniques, provided only a laser source is available. Its sensitivity range spans at least the same range as the Mach–Zehnder interferometer. If

[63] I. I. Komisarova, G. V. Ostrovskaya, L. L. Shapiro, and A. N. Zaidel, *Phys. Lett.* **A29**, 262 (1969).
[64] R. A. Jeffries, *Phys. Fluids* **13**, 210 (1970).

we arbitrarily set fringe shifts between 0.1 and 30 fringes as readily detectable, at the ruby wavelength this encompasses density-path length products [see Eq. (11.3.1)] between 3×10^{16} and $10^{19}/\mathrm{cm}^2$.

The flexibility of holographic systems makes it feasible moreover to extend this range at both limits. For instance, when fringes crowd together the region of interest can be magnified by conventional optics. In a conventional interferometer this would be impractical because of the requirement of phase matching optics in the comparison beam. Also, if two holograms are made simultaneously in different wavelengths instead of sequentially with the same wavelength and the reference beams are incident at appropriately related angles, interference fringes will be formed with a sensitivity to integrated density that is a function of the wavelength difference. (The principle has been demonstrated in holographic contour mapping,[65] but not yet applied to plasmas to our knowledge.) For extending the lower sensitivity limit various multiple path geometries can be used. A method that does not suffer from the spatial resolution loss usually resulting from distinct optical paths through the test section in multiple path applications has been reported by Weigl et al.[66] They used a lossy resonant cavity and the limited laser coherence length to match the reference beam path length to a particular number of traversals of the experimental medium in order to form the hologram, with the remaining multiply reflected beams acting only like an incoherent background fogging. Fringe multiplication up to six has been demonstrated.

11.4.2.2.2.2. CINE-INTERFEROMETRY. Several interferograms during a single plasma discharge are more difficult to obtain with present laser sources than with conventional sources. If, however, the laser source has sufficient duration, multiframe interferograms can be produced holographically by the method of "live" fringes. In this case only the no plasma case is recorded holographically. After processing, the hologram is carefully repositioned. The virtual image reconstruction of the processed hologram is viewed simultaneously with the actual event itself. The reference beam becomes the reconstruction beam and the scene beam illuminates the plasma. The resulting interference pattern from the reconstructed hologram and the live plasma event is photographed with image converter framing cameras. The method was reported by Jahoda.[67] Since that publication greater reconstruction efficiency has been obtained with 10E75 plates, which permitted the use of the same laser for hologram

---

[65] B. P. Hildebrand and K. A. Haines, *J. Opt. Soc. Amer.* **57**, 155 (1967).

[66] F. Weigl, O. M. Friedrich, Jr., and A. A. Dougal, *IEEE J. Quantum Electron.* **5**, 360 (1969).

[67] F. C. Jahoda, *Appl. Phys. Lett.* **14**, 341 (1969).

construction and playback. Sequences of four interferograms with a minimum resolution time of 5 nsec have now been obtained at arbitrary preselected times within a 50 μsec interval. An alternative possibility for multiple frames, unique to holography's ability to store several images on the same hologram emulsion independently by varying the direction or position of the reference beam, is to utilize a single giant laser pulse several times after delaying it by varying time-of-flight paths.[68] Buges[69] has announced a ten-frame camera of 1.5-nsec frame time and 1-nsec interframe separation designed for diagnostics of laser produced plasma.

11.4.2.2.3. NONINTERFEROMETRIC METHODS. The practical success of holographic interferometry has overshadowed other diagnostic applications of holography that are slowly being realized. In most other holographic applications (schlieren, phase contrast, or shadowgraphs), it is necessary to preserve the phase distribution of the plasma for direct observation in the hologram reconstruction. Additional restrictions are then imposed. Obviously, one does not diffuse the scene beam, and every element in the beam path must be of high optical quality in order not to cause phase distortions of the beam. The phase distribution in the reconstructed wavefront will also be somewhat degraded by transmission through the hologram emulsion and substrate itself. Therefore, holographic recording is not desirable where the ultimate in sensitivity is to be achieved. Within these limitations, however, the reconstructed beam can be used in any way that the original beam might have been used. The primary advantage is that instead of the original wavefront occurring just once as a transient pulse, the reconstructed wavefront can be available permanently while the diagnostic technique is adjusted and optimized or even switched from one type of measurement to another.

An interesting variant has been proposed[70,71] in which only the reference beam path must have minimal phase distortion optics. If the processed hologram is accurately repositioned as in the live fringe interferometric method described above, the scene beam of originally complicated phase structure will reconstruct the simple phase structure reference beam except for changes in the scene beam caused by the object under investigation between the time of hologram exposure and the reconstruction. These phase differences in the scene will be superposed on the reference beam structure and can then be studied in, for instance, a conventional schlieren

[68] A. Kakos, G. V. Ostrovskaia, Yu. I. Ostrovsky, and A. N. Zaidel, *Phys. Lett.* **23**, 81 (1966).
[69] J. C. Buges, *C.R. Acad. Sci. Ser. B* **268**, 1624 (1969).
[70] T. Tsuruta and Y. Itoh, *Jap. J. Appl. Phys.* **8**, 96 (1969).
[71] R. E. Brooks, *Appl. Opt.* **8**, 2351 (1969).

arrangement independent of the constant phase distortion (e.g., poor windows) that surround the object of study.

Since the various noninterferometric methods are harder to evaluate quantitatively than interferograms and even then yield density derivatives rather than density directly, they have often been used in plasma diagnostics only because they were experimentally less demanding than interferometry. Because holography inverts the relative ease decidedly in favor of interferometry (permitting diffusers, poor quality windows) there will less often be reason to apply these other methods. Nonetheless they are complementary rather than redundant and will have independent merit in particular cases. Holography will then be particularly useful if ultimate sensitivity is not required and the continued availability of the information bearing beam permits simpler quantitative evaluation, i.e., it will be useful for methods that ordinarily are useful in steady state situations but not with transient events. An example is the use of moiré plates in the reconstructed beam to evaluate angular deviations (density gradients) down to 10^{-4} rad.[72] The choice of ruling spacing, plate separation and angular orientation, all "after the fact," makes quantitative evaluation feasible in any portion of the field. It is also likely that simple shadowgraphs can be made more quantitative through holography by having the whole range of subject-to-screen distances available, particularly when employing magnification methods in the reconstruction. Magnification changes are produced by relocation of the reconstruction point source.

---

[72] F. C. Jahoda, Los Alamos Sci. Lab. Rep., LA–3968–MS. Los Alamos Sci. Lab., Los Alamos, New Mexico, 1968.

12. DEEP SPACE PLASMA MEASUREMENTS*

12.1. Introduction

This chapter describes experimental methods used for the study of plasma in extraterrestrial space by means of instruments carried on earth satellites and space probes. The techniques to be described were developed primarily for observations within the interplanetary medium (the solar wind) (cf. review by Hundhausen[1]) but have also been used for the study of plasma within the magnetosheath and more recently within the magnetotail and the outer magnetosphere[2] (also see review by Gringauz[3]). (Figure 1 illustrates the division of space into distinct regions on the basis of plasma properties and identifies the nomenclature.) This chapter deals only with measurements beyond the plasmasphere. Not included are methods used primarily for observations of ionospheric plasma. The emphasis is on the physical principles underlying the instrumentation and on the methods for analyzing the observations. The specialized technological problems raised by spacecraft operation (such as power and weight limitation, reliability, thermal control, and compatibility with other spacecraft systems) belong to engineering rather than physics and will not be treated here.

12.1.1. Microscopic Character of the Instruments

Space plasma differs in many respects from laboratory plasma and requires correspondingly different measurement techniques. The most obvious is a difference of scale. Table I summarizes some of the properties of plasma observed in various regions of space. Of the various characteristic lengths that can be defined in a plasma, the shortest in practically all explored regions of extraterrestrial space is the Debye length, which nonetheless has a value of the order of or greater than 10 meters. On

---

[1] A. J. Hundhausen, *Space Sci. Rev.* **8**, 690 (1968).
[2] S. J. Bame, *in* "Earth's Particles and Fields" (B. M. McCormac, ed.), p. 359. Reinhold, New York, 1968.
[3] K. I. Gringauz, *Rev. Geophys.* **7**, 339 (1969).

---

* Part 12 by Vytenis M. Vasyliunas.

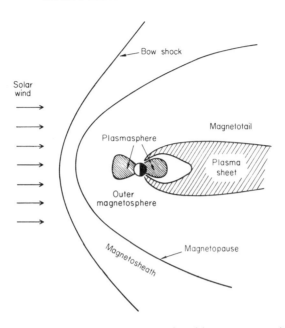

FIG. 1. Sketch of the principal plasma regions found in outer space within $\sim 10^5$ km of the earth. The view is in the noon–midnight meridian plane. The "plasmasphere" is a region of very dense ($\gtrsim 10^3$ particles/cm^3) cold plasma that can be considered an extension of the ionosphere. The earth's magnetic field is effectively confined to a region contained within a surface called the magnetopause. The part of this region lying on the dark side of the earth is usually called the magnetotail or magnetospheric tail and contains a central region of enhanced particle fluxes, the plasma sheet. The magnetopause acts as an obstacle to solar wind flow, which undergoes a sharp shocklike transition (the bow shock) upstream of it. The region between the bow shock and the magnetopause is termed the magnetosheath (an older, now obsolete, term is "transition region"). [For a more detailed description, see, e.g. S. J. Bame, *in* "Earth's Particles and Fields" (B. M. McCormac, ed.), p. 359. Reinhold, New York, 1968, and K. I. Gringauz, *Rev. Geophys.* **7**, 339 (1969), and references therein.]

the other hand, the largest plasma detectors flown have dimensions not exceeding 10 cm and are carried by satellites usually no more than several meters in size (although many satellites have long thin booms that may reach 20–30 meters). Hence the length scale of the instrument is much shorter than any of the characteristic length scales of the plasma, and on this scale the plasma is expected to appear as merely an assemblage of noninteracting charged particles, without the collective behavior generally implied by the word *plasma*. Satellite-borne plasma detectors, accordingly, are not plasma instruments in the sense that, e.g., Langmuir probes are; they are simply devices for the detection and analysis of charged particle beams.

TABLE I. Typical Orders of Magnitude of Plasma Parameters in Various Regions of Outer Space

	Solar wind	Magnetosheath	Outer magnetosphere and plasma sheet
Particle density (1/cm^3)	10	20	1
Temperature (eV)	10	10^2	10^3
Bulk flow speed (km/sec)	4×10^2	2×10^2	10
Magnetic field (gauss)	5×10^{-5}	10^{-4}	2×10^{-4}
Total particle flux (1/cm^2/sec)			
protons	10^8	10^8	4×10^7
electrons	10^9	5×10^9	2×10^9
Length scales, km			
Debye length	10^{-2}	2×10^{-2}	3×10^{-1}
electron gyroradius[a]	2	3	5
proton gyroradius[a]	3×10^3	5×10^3	10^4
scale sizes of typical structures	10^6–10^8	10–10^4	10^4–10^5
mean free path[b]	2×10^7	10^9	2×10^{12}

[a] Based on thermal speed.
[b] For proton–proton Coulomb collisions.

12.1.2. The Problem of Spacecraft–Plasma Interaction

Two further properties of extraterrestrial plasma should be noted from Table I: (a) typical particle energies range from about 10 eV to several kiloelectron volts; (b) the total electron flux generally considerably exceeds the total positive ion flux. From the latter fact one might expect that a spacecraft immersed in such a plasma would quickly acquire a negative charge sufficient to repel most of the electrons in order to have no net current to the spacecraft. This would imply a negative spacecraft potential with respect to the plasma of the order of the mean electron energy per unit charge, large enough to affect seriously any plasma measurements being made. However, several effects exist that may prevent the buildup of such large potentials. The most important arises from the fact that the spacecraft is continually exposed to sunlight; thus photoelectrons are produced at the surface of the spacecraft, with energies of a few electron volts, and their total flux is believed to be of the order of 10^{-8} A/cm^2 ($\sim 10^{11}$ electrons/cm^2/sec),[4] comparable to or larger than the total plasma electron flux. Furthermore, electrons of several hundred electron volts energy striking a metal surface produce secondary electrons, again with energies of a few electron volts; the average number of secondaries

[4] H. E. Hinteregger, K. R. Damon, and L. A. Hall, *J. Geophys. Res.* **64**, 961 (1959).

per incident electron may exceed unity,[5] leading to a net positive current from electron bombardment. The possibility thus exists that the condition of zero net current to the spacecraft is achieved, not by repelling the plasma electron flux, but by balancing it with a large outgoing photoelectron (and perhaps secondary electron) flux; the spacecraft potential then might be only of the order of a few volts and would not appreciably affect the typical plasma measurements at energies of tens of electron volts and higher. To the author's knowledge, a detailed study of the spacecraft-plasma interaction, taking into account the important factors of energetic plasma particle flux, photoelectron and secondary electron emission, does not exist and the aforementioned possibility has been neither justified nor disproved (existing treatments of the problem[6] deal only with low-energy ionospheric plasma and neglect photoelectron effects).

As a practical matter, in all observational work reported to date, it has been assumed that spacecraft potentials do not significantly affect any particles with energies higher than ~ 10 eV, and that fluxes of these particles detected by instruments on the spacecraft accurately represent the fluxes present in the plasma. This assumption is to some extent supported by the actual observations, which show that the plasma electron flux incident on the spacecraft is indeed much larger than the positive ion flux; furthermore, the total number density of the observed electrons is, within errors of measurement, equal to the number density of the observed positive ions,[2,7] as it should be if the measurements indicate conditions in the plasma and are not affected by the spacecraft.

12.2. Instrumentation

12.2.1. General Survey

The first observations of extraterrestrial plasma were made by Gringauz and co-workers[8] by means of very simple probes consisting of a collector plate behind a grid carrying a steady retarding potential (of the order of 10 volts in these measurements). Measurement of the current to the collector indicated the flux of positive ions (or electrons, depending on the polarity of the retarding potential) whose energies per unit charge exceeded the potential on the retarding grid. An additional grid, negatively biased with respect to the collector, was inserted between the collector and the

---

[5] K. G. McKay, *Advan. Electron.* **1,** 65 (1948).

[6] L. W. Parker and B. L. Murphy, *J. Geophys. Res.* **72,** 1631 (1967).

[7] M. D. Montgomery, S. J. Bame, and A. J. Hundhausen, *J. Geophys. Res.* **73,** 4999 (1968).

[8] K. I. Gringauz, V. V. Bezrukikh, V. D. Ozerov, and R. E. Rybchinskii, *Dokl. Akad. Nauk SSSR* **131,** 1301 (1960); *Sov. Phys. Dokl.* **5,** 361 (1961).

retarding grid in order to suppress emission of photoelectrons from the collector; fairly large corrections still had to be made for photoelectron emission from the suppressor grid itself. Instruments of this type have been widely used for measurements of ionospheric plasma[9]; however, they have not been successfully adapted to measure plasma of the kind found in the solar wind and will not be further discussed here.

Two types of instruments have been providing the bulk of deep space plasma observations to date. The first, the modulated-potential Faraday cup, introduced[10] and extensively used by the M.I.T. group, is in many respects similar to the retarding potential probe just discussed but uses an alternating instead of a steady retarding potential to obtain a measurement of the plasma particle energy spectrum. The second selects particles of a specified energy by requiring them to pass between two curved plates with a potential difference between them. An important improvement possible with detectors of the second type is the measurement of flux by counting individual particles rather than measuring currents electrically, thus allowing a considerable increase in sensitivity.

All these instruments can only measure the energy per charge spectrum of the particles. Since hydrogen is expected to be the dominant constituent of extraterrestrial space, it has usually been assumed that the negative particles observed were all electrons and the positive particles predominantly protons (however, significant amounts of alpha particles, which are expected to have the same bulk speed as the protons and hence twice the energy per charge, have always been found in the solar wind). An instrument capable of distinguishing individual ion species has recently been developed by combining a curved-plate analyzer with a crossed electric and magnetic field velocity selector.[11,12]

In this section the general properties of these various types of detectors will be described, in a systematic rather than historical sequence. A list of the spacecraft that have carried plasma detectors appears in Table II, together with references to available descriptions of the instruments.

12.2.2. Energy Measurement

12.2.2.1. The Modulated-Potential Faraday Cup—Basic Principles.

In its simplest form the modulated-potential Faraday cup, shown schemati-

---

[9] W. C. Knudsen, *J. Geophys. Res.* **71,** 4669 (1966).

[10] H. S. Bridge, C. Dilworth, B. Rossi, F. Scherb, and E. F. Lyon, *J. Geophys. Res.* **65,** 3053 (1960).

[11] K. W. Ogilvie, N. McIlwraith, and T. D. Wilkerson, *Rev. Sci. Instrum.* **39,** 441 (1968).

[12] K. W. Ogilvie, R. I. Kittredge, and T. D. Wilkerson, *Rev. Sci. Instrum.* **39,** 459 (1968).

TABLE II. Deep Space Plasma Experiments, 1959–1968[a]

Spacecraft	Year of launch	Type of detector[b]	Experimenters[c] and references[d]
Lunik 1	1959	RPP	Acad. Sci. U.S.S.R.[e]
Lunik 2	1959	RPP	Acad. Sci. U.S.S.R.
Lunik 3	1959	RPP	Acad. Sci. U.S.S.R.
Venus 1	1961	RPP	Acad. Sci. U.S.S.R.
Explorer 10	1961	MFC	M.I.T.[f]
Explorer 12	1961	CPA	Ames[g]
Mariner 2	1962	CPA	J.P.L.[h]
Explorer 14	1962	CPA	Ames
Mars 1	1962	RPP	Acad. Sci. U.S.S.R.
Explorer 18 (IMP 1)	1963	MFC	M.I.T.[i]
		CPA	Ames[j]
Electron 2	1964	RPP	Acad. Sci. U.S.S.R.
		CPA	Acad. Sci. U.S.S.R.[k]
Vela 2A, 2B	1964	CPA	Los Alamos[l]
OGO 1	1964	MFC	M.I.T.[m]
		CPA	Ames
Explorer 21 (IMP 2)	1964	MFC	M.I.T.
		CPA	Ames
Mariner 4	1964	MFC	M.I.T.–J.P.L.[n]
Zond 2	1964	RPP	Acad. Sci. U.S.S.R.
Explorer 28 (IMP 3)	1965	MFC	M.I.T.
		CPA	Ames
Vela 3A, 3B	1965	CPA	Los Alamos
Venus 2	1965	RPP	Acad. Sci. U.S.S.R.
Venus 3	1965	RPP	Acad. Sci. U.S.S.R.
Pioneer 6	1965	MFC	M.I.T.[o]
		CPA	Ames[p]
Luna 10	1966	RPP	Acad. Sci. U.S.S.R.
OGO 3	1966	MFC	M.I.T.
		CPA	Ames
		CPA	Iowa[q]
Explorer 33	1966	MFC	M.I.T.[r]
Pioneer 7	1966	MFC	M.I.T.
		CPA	Ames
ATS 1	1966	MFC	Rice[s]
Explorer 34 (IMP F)	1967	CPA-VS	GSFC–Maryland[t]
		CPA	Iowa
Venus 4	1967	RPP	Acad. Sci. U.S.S.R.
Mariner 5	1967	MFC	M.I.T.–J.P.L.
Explorer 35	1967	MFC	M.I.T.
Vela 4A, 4B	1967	CPA	Los Alamos[u]
Pioneer 8	1967	CPA	Ames
OGO 5	1968	CPA	Iowa
		MFC, CPA	J.P.L.

* See p. 55 for footnotes to table.

12.2. INSTRUMENTATION

cally in Fig. 2, consists of a cylindrical metal cup open at one end with a collector plate at the other end and a minimum of four planar grids. Grid 1 is usually grounded (throughout this chapter "ground" means the potential of the skin of the spacecraft which, as discussed previously, is assumed to differ by no more than some few volts from the potential of the surrounding plasma); it and the walls of the cup, also grounded, form a closed equipotential surface around the detector and prevent any electric fields generated within the detector from penetrating outside. Grid 2 is

---

[a] Only instruments intended for the study of plasma in the outer magnetosphere and beyond are included. The list may not be complete.

[b] RPP is the retarding potential (dc) probe; MFC is the modulated Faraday cup; CPA the curved-plate analyzer; VS the velocity selector.

[c] Acad. Sci. U.S.S.R. is the Academy of Sciences of the U.S.S.R.; M.I.T. is the Massachusetts Institute of Technology; Ames is the NASA Ames Research Center; J.P.L. is the Jet Propulsion Laboratory, California Institute of Technology; Los Alamos is the Los Alamos Scientific Laboratory, University of California; Iowa is the University of Iowa; Rice is the Rice University; GSFC–Maryland is the NASA Goddard Space Flight Center and University of Maryland.

[d] Only published papers containing descriptions of instrumentation are included. Essentially identical instruments have often been flown on several spacecraft; in such cases only the first one has been referenced.

[e] K. I. Gringauz, V. V. Bezrukikh, V. D. Ozerov, and R. E. Rybchinskii, *Dokl. Akad. Nauk SSR* **131**, 1301 (1960); *Sov. Phys. Dokl.* **5**, 361 (1961).

[f] A. Bonetti, H. S. Bridge, A. J. Lazarus, B. Rossi, and F. Scherb, *J. Geophs. Res.* **68**, 4017 (1963).

[g] M. Bader, *J. Geophys. Res.* **67**, 5007 (1962).

[h] M. Neugebauer and C. W. Snyder, *J. Geophys. Res.* **71**, 4469 (1966).

[i] H. S. Bridge, A. Egidi, A. Lazarus, E. Lyon, and L. Jacobson, *Space Res.* **5**, 969 (1965).

[j] J. H. Wolfe, R. W. Silva, and M. A. Myers, *J. Geophys. Res.* **71**, 1319 (1966).

[k] S. N. Vernov, V. V. Melnikov, I. A. Savenko, and B. I. Savin, *Space Res.* **6**, 746 (1966).

[l] S. Singer, *Proc. IEEE* **53**, 1935 (1965).

[m] V. M. Vasyliunas, *J. Geophys. Res.* **73**, 2839 (1968).

[n] A. J. Lazarus, H. S. Bridge, J. M. Davis, and C. W. Snyder, *Space Res.* **7**, 1296 (1967).

[o] A. J. Lazarus, H. S. Bridge, and J. Davis, *J. Geophys. Res.* **71**, 3787 (1966).

[p] J. H. Wolfe, R. W. Silva, D. D. McKibbin, and R. H. Mason, *J. Geophys. Res.* **71**, 3329 (1966).

[q] L. A. Frank, *J. Geophys. Res.* **72**, 185 (1967).

[r] E. F. Lyon, H. S. Bridge, and J. H. Binsack, *J. Geophys. Res.* **72**, 6113 (1967).

[s] J. W. Freeman, Jr., *J. Geophys. Res.* **73**, 4151 (1968).

[t] K. W. Ogilvie, N. McIlwraith, and T. D. Wilkerson, *Rev. Sci. Instrum.* **39**, 441 (1968).

[u] M. M. Montgomery, S. J. Bame, and A. J. Hundhausen, *J. Geophys. Res.* **73**, 4999 (1968).

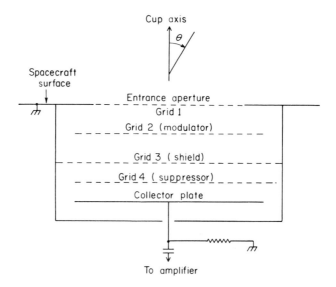

Fig. 2. Schematic diagram of the modulated-potential Faraday cup.

the modulator grid; an alternating potential, usually in the form of a square wave at a frequency of a few kilohertz, is applied to it. Grid 3, the shield grid, is grounded and prevents the modulator potential from inducing currents on the collector by direct capacitive coupling. Grid 4 is maintained at a steady negative potential (typical values range from -50 to -150 volts) and serves to suppress the emission of photoelectrons and/or secondary electrons from the collector plate.

A typical instrument may have a diameter of ~ 10 cm and a depth of ~ 5 cm. The shield grid is usually a phosphor-bronze grid with an optical transparency of $\sim 33\%$ at normal incidence; the others are $\sim 90\%$ transparent tungsten grids.

To describe the functioning of the instrument, let V_1 and V_2 be the limits of the square wave modulator potential; $V_2 > V_1$ and either $V_1 > 0$ or $V_2 < 0$, depending on whether positive or negative particles are to be measured. Consider a particle of charge q and kinetic energy E incident at an angle θ to the axis of the cup. Then if

$$E \cos^2 \theta < qV_1, \qquad (12.2.1)$$

the particle is reflected by the modulator grid at all times. If

$$qV_1 < E \cos^2 \theta < qV_2, \qquad (12.2.2)$$

the particle can reach the collector during the lower half of the modulation cycle (modulator potential is V_1) and is reflected by the modulator grid during the upper half (V_2), thus producing an alternating current at the collector that is at the same frequency as the modulator potential and is 180° out of phase. Finally, if

$$qV_2 < E \cos^2 \theta \qquad (12.2.3)$$

(this includes particles of opposite polarity to the modulator potential), the particle can reach the collector at all times and produces only a steady current, to first approximation (corrections to this are discussed further on). (Negative particles for which $E \cos^2 \theta/q < V_s$, the potential on the suppressor grid, never reach the collector regardless of the modulator potential.)

The current to the collector thus contains two components: an alternating current proportional to the flux of incident particles satisfying Eq. (12.2.2), and a steady current due to other particles. The alternating current is measured and provides information on particles within the "energy window" defined by (12.2.2). By varying the limits of the modulating potential, particles in different "energy windows" can be observed.

12.2.2.1.1. HIGHER-ORDER CORRECTIONS. The distance between adjacent grids is usually made small compared to the diameter of the cup; hence the electric fields in the center of the cup can be calculated to a good approximation by treating the grids as infinite equipotential planes. An incident charged particle with sufficient energy to reach the collector (i.e., satisfying Eq. (12.2.3)) may nevertheless be affected by the modulator: if it is incident at a nonzero angle to the cup axis, it will be laterally displaced in passing through the electric fields of the modulator by an amount ΔL that is readily shown to be

$$\Delta L = L \tan \theta \, \frac{1 - (1 - x)^{1/2}}{1 + (1 - x)^{1/2}}, \qquad (12.2.4)$$

$$x \equiv qV/E \cos^2 \theta,$$

where V is the potential on the modulator and L is the distance between the two grounded grids surrounding the modulator (i.e., grids 1 and 3). The largest possible displacement, which occurs when either $x \to 1$ or $x \to -\infty$, has the magnitude L. If a particle reaches the collector plate near its edge, the difference between the lateral displacements at the upper and lower values of the modulator potential may be large enough so that the particle strikes the collector during one but misses it during the other, thus giving rise to an alternating current (it can be easily seen that this current is also 180° out of phase with the modulator potential). Thus the alternating current measures not only the flux of particles satisfying

Eq. (12.2.2), i.e., lying within the nominal energy window of the detector, but also some fraction of the flux of particles satisfying Eq. (12.2.3), i.e., lying above the nominal energy window or having the wrong polarity. It will shortly appear that this fraction is usually of the order of a few percent; hence this effect can be neglected unless the flux of particles within the nominal energy window is more than an order of magnitude smaller than the flux above or the flux of particles of opposite polarity, a situation which is not common but has occasionally been found to occur.[13,14]

12.2.2.1.2. ANGULAR RESPONSE. The most fundamental description of any detector is provided by the specification of its transmission function as a function of energy and direction of incidence. The transmission function (also known as the response function) is defined as the fraction of a beam of particles (of a given energy and direction, incident uniformly upon the entrance aperture) that is detected by the instrument; thus the measured current produced by a beam is equal to the flux density of the beam, times the area of the entrance aperture, times the charge of a single particle, times the transmission function.

In a modulated-potential Faraday cup the transmission function is determined primarily by (a) the modulating potential, which selects particles within a given range of velocities along the axis of the cup, as described earlier, and (b) the geometrical construction of the cup, which limits the range of angles from which particles can be accepted. The transparency of the grids and its variation with angle must also be taken into account. Figure 3 shows, as an example, the calculated transmission function of a Faraday cup flown on the IMP 2 satellite. As can be seen, the transmission function at a given angle is approximately constant over the energy range specified by Eq. (12.2.2) and at angles other than zero has a much smaller high energy "tail" as well as some response to particles of negative charge. At a given value of $E \cos^2 \theta$ the transmission function decreases uniformly with increasing angle and falls to zero, for typical cups, at an angle from the axis ranging from 45° to 60°.

The cylindrically symmetric construction of the usual Faraday cup implies that its transmission function depends only on energy and the polar angle with respect to the direction along the axis of the cup. As a result of this axial symmetry, the measurements provide no information on the azimuthal angle of incidence of the plasma. This limitation can be

[13] J. H. Binsack, *in* "Physics of the Magnetosphere" (R. L. Carovillano, J. F. McClay, and H. R. Radoski, eds.), p. 605. Reidel, Dordrecht, Holland, 1969.

[14] S. Olbert, *in* "Physics of the Magnetosphere" (R. L. Carovillano, J. F. McClay, and H. R. Radoski, eds.), p. 641. Reidel, Dordrecht, Holland, 1969.

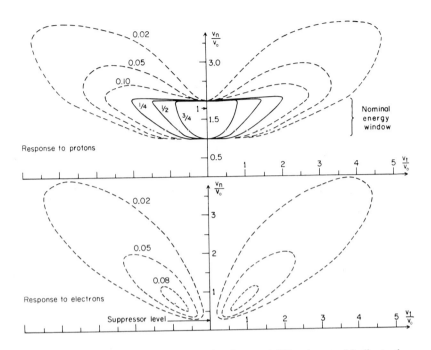

FIG. 3. Transmission function of a modulated-potential Faraday cup (similar to the one flown on IMP 2) operated in the positive ion mode. The transmission function, here and in subsequent figures, has been normalized so that its maximum value is unity. Shown are contours in velocity space of constant value of the normalized transmission function. The top figure is the response to positive particles, the bottom to negative particles, the detector in each case being operated in the positive ion mode. V_n is the velocity component along the normal to the entrance aperture, V_t the component transverse to it, and V_0 the speed corresponding to the bottom of the nominal energy window. The transmission function has rotational symmetry about the V_n axis. Dashed contours represent the 10% or less response due to the lateral displacements discussed in the text.

overcome to some extent by splitting the collector plate into several segments (usually two or three) and measuring the current from each segment separately. Figure 4 illustrates such a split-collector cup, used on the OGO satellites, and shows the dependence of its transmission function on direction. In some Faraday cups a "Venetian blind" collimator is placed in front of the entrance aperture in order to narrow down the angular opening of the transmission function to $\gtrsim 20°$ in the plane perpendicular to the collimator plates (but leaving the usual ~50° width in the plane containing the plates).[15]

[15] A. J. Lazarus, H. S. Bridge, and J. Davis, *J. Geophys. Res.* **71**, 3787 (1966).

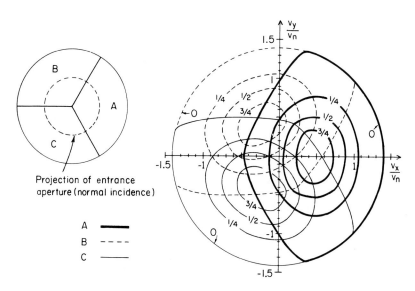

Fig. 4. (Left) A Faraday cup collector plate split into three segments (labeled A, B, C, respectively) to obtain some data on the azimuthal distribution of particle directions. (Right) Dependence of the normalized transmission function for the three collector segments on velocity transverse to the normal incidence direction. V_n is the velocity component along the normal to the entrance aperture and V_x, V_y are the transverse velocity components along axes x, y oriented relative to the collector segments as shown.

12.2.2.1.3. Extraneous Effects. From extensive experience with modulated Faraday cups, both in laboratory testing and in spacecraft observation, it appears that when the cup is operated in the positive ion mode (i.e., with positive modulating potentials), its actual transmission function is adequately represented by the theoretical transmission function (such as that shown in Fig. 3) calculated from particle trajectories and cup geometry. In particular, the detector in this mode has no appreciable response to sunlight.[16] From the configuration of the potentials on the various grids, it is easily seen that only photoelectrons released from the suppressor grid can reach the collector; they, however, are not modulated and produce only an unmeasured steady current (historically this elimination of photocurrent effects from the suppressor grid was one of the principal reasons why the modulated-potential Faraday cup was developed[10]).

On the other hand, when the detector is operated in the electron mode (with negative modulating potentials), several additional sources of

---

[16] A. Bonetti, H. S. Bridge, A. J. Lazarus, B. Rossi, and F. Scherb, *J. Geophys. Res.* **68**, 4017 (1963).

alternating currents not included in the theoretical transmission function may be present. In this mode the detector is strongly sensitive to sunlight, which produces alternating currents as large as or larger than typical electron currents to be measured; successful observation of electrons has so far only been possible with the detector facing well away from the sun.[17] The source of these modulated photoelectron signals is not well understood, but they are thought to be produced by photoelectrons released from the modulator grid. Two mechanisms by which these photoelectrons can be modulated have been proposed: (a) the fraction of photoelectrons moving toward the collector (rather than out the entrance aperture) may be affected by the strong inhomogeneous electric fields immediately adjacent to the grid wires, which vary with the modulator potential; (b) the photoelectrons strike the suppressor grid with an energy per charge essentially equal to the difference between the suppressor and modulator potentials (thus varying periodically with the modulator potential) and produce a secondary electron flux that is modulated because the secondary electron yield (number of secondaries per primary) varies with the primary electron energy.

Furthermore, in the electron mode the detector has a sensitivity to protons that arises because the protons strike the modulator grid at an energy/charge that is the sum of the modulator potential and the proton energy/charge outside the detector; as in (b) above, this gives rise to a modulated secondary electron current.[17,18] Since the secondary electron yield for protons striking a metal surface is approximately linear with proton energy (up to energies of ~ 10 keV),[19] the modulated secondary electron flux is proportional to the difference between the upper and lower limits of the modulator potential and is independent of the proton energy; it is also in phase with the modulation rather than 180° out of phase, allowing the occurrence of the effect to be recognized. The magnitude of this modulated current is approximately one percent of the incident proton current for a modulation potential range of 1 kV,[17] and in practice it is important only within regions having a very high proton density, such as the plasmasphere (refer to Fig. 1). The effect could be suppressed by placing in front of the modulator another grid with a steady positive potential large enough to repel the incoming protons.

Finally, incoming electrons detected by the cup produce secondary electrons at both the modulator and suppressor grids; the secondary flux is proportional to the primary flux and hence is modulated along with

---

[17] V. M. Vasyliunas, *J. Geophys. Res.* **73**, 2839 (1968).
[18] J. H. Binsack, *J. Geophys. Res.* **72**, 5231 (1967).
[19] D. B. Medved and Y. E. Strausser, *Advan. Electron. Electron Phys.* **21**, 101 (1965).

it. The main effect is simply to enhance the incident electron current by an amount ranging up to 10%, depending on the electron energy.

12.2.2.2. The Curved-Plate Analyzer—General Description. The curved-plate electrostatic analyzer was developed as a laboratory instrument long before the era of space exploration (see references in the work of Paolini and Theodoridis[20]). Its general form is shown schematically in Fig. 5. It consists of two parallel curved plates, which are segments either

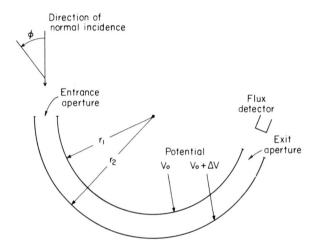

Fig. 5. Schematic diagram of a curved-plate analyzer.

of a cylinder or of a sphere, with a potential difference between them (in typical space applications, the radii of curvature of the plates are of the order of 10 cm and the separation between the plates is of the order of 1 cm or less). A charged particle entering the entrance aperture at normal incidence will move in a circular orbit of radius r if its kinetic energy E (when inside the analyzer), charge q, the potential difference between the plates ΔV, and the radii of curvature of the inner and outer plates, r_1 and r_2, are related by the equation

$$2E = q\,\Delta V/\log(r_2/r_1) \qquad (12.2.5)$$

for a cylindrical analyzer, or

$$2E = q\,\Delta V/r\left(\frac{1}{r_1} - \frac{1}{r_2}\right) \qquad (12.2.5a)$$

[20] F. R. Paolini and G. C. Theodoridis, *Rev. Sci. Instrum.* **38**, 579 (1967).

for a spherical analyzer (evidently $r_1 < r < r_2$). If $r_1 \equiv r_0 - \frac{1}{2}\Delta r$, and $r_2 \equiv r_0 + \frac{1}{2}\Delta r$, then to lowest order in $\Delta r/r_0 \ll 1$ both equations reduce to

$$2E \simeq q\,\Delta V r_0/\Delta r. \tag{12.2.5b}$$

Particles which are incident at a direction sufficiently different from normal or whose energies differ sufficiently from the value given by Eq. (12.2.5) will strike one or the other of the plates before reaching the exit aperture. Thus the flux of particles out of the exit aperture corresponds to the flux of incident particles within a narrow range of energies and directions (in the plane perpendicular to the plates). This steady flux of particles can be detected either by measuring the current to a collector plate or by counting pulses from a single particle detector.

It should be noted that the kinetic energy E in Eqs. (12.2.5) is the energy the particle has while inside the analyzer. Unless the particle crosses the entrance aperture at the unique point (if it exists) whose potential with respect to points far outside is zero, E is not equal to the particle's energy outside the analyzer. The change in energy is brought about by acceleration in the fringing fields at the entrance aperture, which thus necessarily play a role in determining the transmission function of the analyzer; in particular, from Eq. (12.2.5b) it at once follows that, if the width of the entrance aperture is equal to the separation of the plates (as it usually is in space applications), the range of (outside) particle energies accepted by the analyzer is at least equal to $q\,\Delta V$.

12.2.2.2.1. SURVEY OF VARIOUS GEOMETRIES. The principal properties characterizing a particular curved-plate analyzer are:

(1) Type of plates (spherical or cylindrical).

(2) Ratio of plate spacing to mean radius of curvature ($\Delta r/r_0$ in Eq. (12.2.5b); values ranging from 0.03 to 0.1 have been used in space work.

(3) Length of the plates, usually expressed as an angle (180° corresponds to half a sphere or cylinder, etc.).

(4) Potential of the plates with respect to distant points. A "balanced" analyzer has a potential $-\frac{1}{2}\Delta V$ on the inner plate and $+\frac{1}{2}\Delta V$ on the outer plate, to look at positive ions (the reverse to look at electrons), so that zero potential is midway between the plates. Any other arrangement gives an "unbalanced" analyzer. The main advantage of balanced analyzers is that they minimize the fringing field effects mentioned earlier; on the other hand, they can measure particles of a given polarity only (plate potentials must be interchanged to measure particles of the other polarity). The one type of unbalanced analyzer in spacecraft use [see (d) following] has three plates, the middle one at a potential $+\Delta V$ and the outer two

grounded; the inner pair of plates thus forms an analyzer for protons and the outer pair for electrons.

By varying these properties a large number of different analyzer designs can be obtained. Four main types have been used on spacecraft to date:

(a) Cylindrical plates, 120° in length, balanced. Example: Mariner 2 (see Table II).

(b) Hemispherical: 180°-long spherical plates, balanced (e.g., Vela satellites).

(c) Quadrispherical: 90°-long spherical plates, balanced (e.g., Pioneer 6, 7, 8).

(d) Cylindrical 43° plates, unbalanced three plate arrangement described above (e.g., OGO 3, 5); this is the "Low Energy Proton and Electron Differential Energy Analyzer" or LEPEDEA, developed at the University of Iowa.[21,22] In later versions of this instrument,[23] the plate potential is not steady but is modulated at a frequency of 2.4 kHz by a triangular waveform having an amplitude of 25% of the mean value; since the modulation period is long compared to the transit time of the particles through the analyzer yet short compared to the integration time of the particle detector at the exit aperture, the result is simply to broaden the energy response of the instrument by 25% over that obtained with a fixed plate potential.

Most curved plate analyzers accept particles within a narrow range of angles in the plane perpendicular to the plates (azimuthal angles) but over a wide range in the plane tangent to the plates at the entrance aperture (polar angles). Quadrispherical analyzers [type (c) above] developed at the NASA Ames Research Center were designed to measure simultaneously the particle flux as a function of polar angle, by means of the scheme illustrated in Fig. 6.[24] A relatively narrow entrance aperture is used, but a wide collector plate is placed at the exit aperture and split into a number of separate segments; the current from each segment corresponds to the flux of particles entering the analyzer within a particular narrow range of polar angles. The quadrispherical geometry is essential for this measurement; in a hemispherical analyzer, for instance, particles entering at a particular point on the aperture but at different polar angles will all converge again to a single point after traversing the half sphere.

---

[21] L. A. Frank, Univ. of Iowa Rep. 65–22. Univ. of Iowa, Iowa City, Iowa, 1965.

[22] L. A. Frank, *J. Geophys. Res.* **72,** 185 (1967).

[23] L. A. Frank, W. W. Stanley, R. H. Gabel, D. C. Enemark, R. F. Randall, and N. K. Henderson, Univ. of Iowa Rep. 66–31. Univ. of Iowa, Iowa City, Iowa, 1966.

[24] J. H. Wolfe, R. W. Silva, D. D. McKibbin, and R. H. Mason, *J. Geophys. Res.* **71,** 3329 (1966).

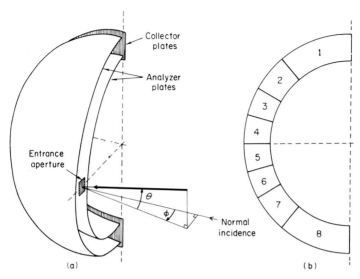

FIG. 6. Schematic diagram of (a) a quadrispherical curved-plate analyzer with (b) a multi-segment collector (face-on view) designed to measure the particle distribution as a function of polar angle θ.

12.2.2.2.2.2. THE COUPLED ENERGY-ANGULAR RESPONSE. Figure 7 illustrates the transmission functions of several curved-plate analyzers used for solar wind observations. An intrinsic feature of all curved-plate analyzers, readily apparent in Fig. 7, is the energy-angle asymmetry or "skewing": the range of energy of particles accepted by the analyzer is a strong function of azimuthal angle ϕ, moving upward with increasing angle (in the sense indicated in Fig. 5); the transmission function appears markedly "tilted" from the normal incidence direction in velocity space. The effect arises because particles entering the analyzer with their velocity vectors directed toward the inner plate must have a higher energy to be transmitted than particles with their velocity vectors directed toward the outer plate. The magnitude of the "skewing" increases with increasing ratio of plate spacing to radius and is more pronounced for a quadrispherical than for a hemispherical analyzer. The existence of this strong coupling between energy and angle required for acceptance implies that a curved-plate analyzer cannot in general be characterized merely by an "energy bandpass" and "angle of acceptance."

If the only potentials in the analyzer are $\pm\frac{1}{2}\Delta V$ on the two plates or ΔV on one plate and ground on the other, it is readily shown by scaling the equations of motion that the transmission function depends only on angles and on the ratio $E/q\,\Delta V$, i.e., the energy dependence scales with the plate potential. This implies that, if at a particular angle particles with

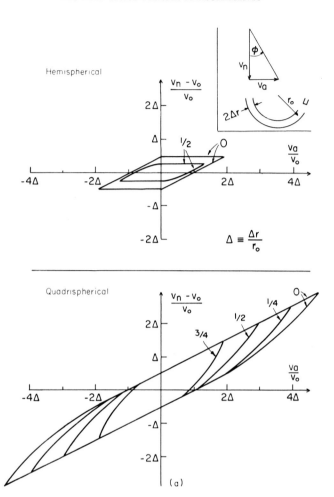

FIG. 7a. Normalized transmission functions of a hemispherical (top) and quadrispherical (bottom) curved-plate analyzer; dependence on normal and azimuthal velocity components (defined as shown in insert). V_0 is the speed corresponding to the energy defined by Eq. (12.2.5b). (From theoretical calculations, kindly provided by A. J. Hundhausen.)

energies between, say $E - \frac{1}{2}\Delta E$ and $E + \frac{1}{2}\Delta E$ are accepted, then $\Delta E/E$ is a constant, independent of the plate potential and depending only on the geometry of the analyzer (it can be shown that $\Delta E/E$ is proportional to $\Delta r/r_0$ if it is small). This result is unaffected by the presence of fringing fields; any departures from it imply that either some potential other than the ΔV on the plates is present (for example, postaccelerating

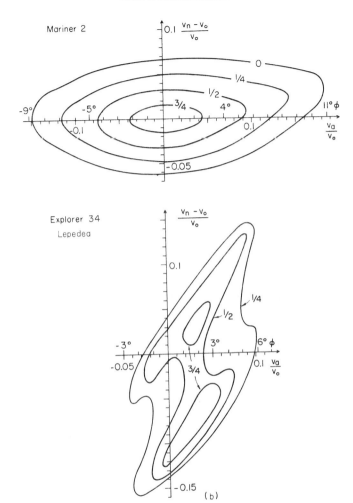

FIG. 7b. Normalized transmission functions of cylindrical curved-plate analyzers; dependence on normal and azimuthal velocity components. (Top) Response of Mariner 2 analyzer [constructed from data by M. Neugebauer and C. W. Snyder, *J. Geophys. Res.* **71**, 4469 (1966)]. (Bottom) Response of channel 6B of LEPEDEA analyzer with modulated plate voltage flown on Explorer 34 (constructed from laboratory calibration curves given by L. A. Frank, W. W. Stanley, R. H. Gabel, D. C. Enemark, R. F. Randall, and N. K. Henderson, Univ. of Iowa Rep. 66–31. Univ. of Iowa, Iowa City, Iowa, 1966.) V_0 corresponds to an energy of 1 keV.

potentials after the exit aperture) or that effects not included in the equations of classical mechanics (e.g., secondary emission or scattering of particles from plates) are occurring.

12.2.2.2.3. EXTRANEOUS EFFECTS. Many of the instruments used to detect particles coming out of the exit aperture are sensitive to sunlight. Because of the curvature of the plates, light cannot reach the detector directly, but in practice it might reach the detector by scattering from the plates; to prevent this, the inside surfaces of the analyzer are usually blackened. It is also possible to construct that portion of the plates struck by direct sunlight out of a highly transparent grid (instead of a solid metal) and to place a light trap behind it.[11]

Two other effects that may be important, especially in analyzers used to observe electrons, are the scattering of electrons from the plates and the production of secondary electrons near the end of the plates. In some instruments the plates are machined with narrow (approximately 0.1 cm) serrations to suppress scattering (and also reflection of light). Secondary electron emission may be reduced or suppressed by placing a negative grid after the exit aperture, or some other negative electrode (e.g., a negative ring in the Mariner 2 analyzer) that can prevent low-energy electrons released at the plates from reaching the particle detector. Alternatively, electron scattering and secondary production may be allowed but their effect estimated from laboratory calibration and included in the specification of the transmission function.

12.2.2.3. The Crossed-Field Velocity Selector. An instrument capable of measuring separately the energy spectra of protons and alpha particles has been developed by the Goddard–University of Maryland group.[11,12] A cylindrical curved-plate analyzer (similar to those described in the previous section) selects ions within a narrow range of energy per charge; particles leaving the exit aperture of the analyzer then enter a crossed electric and magnetic field velocity selector (Wien filter) tuned to a velocity corresponding to the energy per charge of the ion species being observed. With the instrument set to observe alpha particles, the proton contamination is less than 0.3%, and vice versa. A detailed description of the velocity selector, including a discussion of the design requirements imposed by spacecraft use (chiefly the necessity for a permanent magnet and a severe restriction on the total length of the device) has been published.[12]

12.2.2.4. Comparison of Instruments

12.2.2.4.1. ENERGY RANGE. An upper limit to the energy of particles that can be detected is set by the highest available voltage. Voltages up to ~10 kV have been successfully attained in detectors flown to date. Hence a modulated-potential Faraday cup, which requires a voltage as large as the energy per charge of the particles being observed, is at present limited to protons and electrons having energies under 10 keV; this is generally adequate for observations within the solar wind but not within some regions of the outer magnetsphere. On the other hand, a curved-plate analyzer requires a plate voltage which, in typical space instruments, is of the order of 10% of the particle energy per charge; thus it can observe particles up to 50–100 keV, overlapping the energy range accessible to high-energy instruments such as the thin-window Geiger tube (which can detect electrons above ~ 40 keV) (see, e.g., Frank and Van Allen[25]) or the scintillation counter (electrons above ~ 20 keV, protons above ~ 100 keV) (see, e.g., Davis and Williamson[26]).

12.2.2.2.4.2. DIFFERENTIAL VERSUS INTEGRAL MEASUREMENTS. The curved-plate analyzer is an intrinsically differential instrument, measuring particles within a narrow range of energies and angles. A very narrow response enables one to study in some detail the velocity space structure of very narrow particle distributions such as are found in the solar wind. Since the location of such sharp peaks in velocity space is not, however, *a priori* known and may be variable, the narrow energy-angle ranges sampled by the various detector channels must be contiguous and hence the required number of channels is quite large; if there are appreciable gaps in either energy or angle coverage, a significant fraction of the particle distribution may not be observed by the detector, with consequent large errors in estimates of total particle flux. On the other hand, modulated Faraday cups used to date have had relatively wide energy and angular windows (it should be understood, however, that this is not necessarily an intrinsic characteristic of this detector; energy windows as narrow as 2% are to be used in some future instruments,[27] and in principle the angular response could be narrowed down to a few degrees with collimators). The wide response makes it possible to cover a large part of velocity space with a relatively small number of contiguous windows and to obtain reliable total flux measurements no matter how narrow the particle distribution. The question of differential versus integral functioning of detectors is further discussed in Section 12.3.2.

It should also be noted that the ratio $\Delta E/E$ discussed earlier is generally fixed by the geometry in a curved-plate analyzer, whereas it can be electronically varied in a modulated-potential Faraday cup by varying the width of the modulator voltage relative to the mean level.

12.2.2.2.4.3. SENSITIVITY. The entrance aperture of a Faraday cup can be made quite large, since its maximum diameter is limited only by the size of the entire instrument. The curved-plate analyzer, on the other hand, requires a plate separation small compared to their radius and hence has a

[25] L. A. Frank and J. A. Van Allen, *J. Geophys. Res.* **68**, 1203 (1963).
[26] L. R. Davis and J. M. Williamson, *Space Res.* **3**, 365 (1963).
[27] J. Binsack, Private communication, 1968.

rather small entrance aperture. The effective collecting area of a typical Faraday cup is thus at least two orders of magnitude larger than that of a typical curved-plate analyzer; this is somewhat offset by the greater sensitivity of flux measurement methods used with curved-plate analyzers (see Section 12.3). Taking into account both the collecting area and the flux detection sensitivity, the minimum detectable particle flux density is of the same order of magnitude both for a Faraday cup and for a curved-plate analyzer that uses current collector plates for particle detection. The introduction of particle-counting methods in curved-plate analyzers allowed a considerable lowering of the minimum detectable flux level for a given collecting area or, alternatively, a reduction of collecting area for a given flux level. Initially the second alternative was taken: plate separation (and hence the collecting area) was reduced in order to obtain an increase in resolution,[1] without a major gain in sensitivity. In analyzers of the LEPEDEA class, however, particle counting was combined with a relatively wide plate spacing to obtain a minimum detectable flux level two orders of magnitude below that of previous instruments[22] (although at the expense of a greatly enhanced "skewing" of the transmission function).

12.2.2.4.4. ABSOLUTE CALIBRATION. The relatively simple construction of the Faraday cup allows a reliable theoretical calculation of its transmission function, which can be verified by simple laboratory tests; hence the absolute efficiency for the detection of positive ions is known with great certainty. On the other hand, the complicated response of the curved-plate analyzer and the unavoidable presence of significant fringing fields make it difficult to obtain theoretically a useful transmission function, which must then be determined by laboratory calibration (discussed further in Section 12.2.4). The calibration must be detailed enough to determine the complete shape of the transmission function, including the energy-angle "skewing" discussed earlier; knowledge of merely an "energy width" and "angular width" is not adequate. In the absence of such a calibration, the absolute efficiency of the analyzer may be rather uncertain.

On the other side of the ledger, a Faraday cup operated in the positive ion mode has a slight sensitivity to electrons (already discussed in Section 12.2.2.1.1), which under some conditions is significant and must be taken into account[13,14] (in the solar wind and the magnetosheath, however, it can be eliminated almost entirely by setting the negative suppressor grid potential at a value $\gtrsim 100$ volts, large enough to prevent most of the plasma electrons from reaching the collector); such "contamination" is not present in a curved-plate analyzer.

12.2.3. Flux Measurement

In this section we briefly describe the principal methods used to measure the flux of particles striking the collector plate or leaving the exit aperture of the instruments described. Only the basic features will be treated, not the detailed electronic circuitry (detailed descriptions of the electronics for several of the instruments are available as technical reports[28-30]).

12.2.3.1. Direct Current Methods. Curved-plate analyzers that use collector plates as detecting elements measure the direct current from the plates by means of electrometers. Typical instruments have a minimum detectable current of 10^{-14}–10^{-13} A, a minimum integration time of the order of 0.1 sec, and a dynamic range varying from 4 to 7 decades. The measurement is transmitted to a tracking station either in digital form (generally as a binary number of 6–8 bits) or in analog form (generally with at most a few percent precision) depending on the spacecraft telemetry system. Since neither method allows the transmission of a number with a 4-decade range and since the construction of a signal processing system linear over several decades is very difficult, the value of the collector current is not itself the telemetered quantity; instead, the electrometer is designed so that its output is proportional to the logarithm of the input current.

12.2.3.2. Alternating Current Methods. The measured output of a modulated-potential Faraday cup is an alternating current of known frequency and phase. The measurement is accomplished by coupling the collector plate to the signal processing system through a capacitor, then passing the ac signal through a preamplifier, through a narrow-band filter tuned to the known frequency, and through a compression amplifier that produces an output approximately proportional to the logarithm of the input current (for the same reasons as in the case of the dc electrometers discussed above). The ac signal finally must be rectified; in the early instruments this was accomplished with a simple full-wave rectifier circuit, but at present a synchronous detector is generally used to exploit the fact that the phase of the signal is known (this results in a very narrow bandwidth and a consequent reduction of the noise level). Typical integration times, in the instruments flown to date, are of the order of tens of milliseconds (considerably shorter than those of the dc electrometers).

---

[28] E. F. Lyon, Lincoln Lab. Rep. 52G–0017. Lincoln Lab., M.I.T., Cambridge, Massachusetts, 1961.

[29] C. Josias and J. L. Lawrence, Jr., Tech. Rep. 32–492. Jet Propulsion Lab., Pasadena, California. 1964.

[30] Final Engineering Report for the M.I.T. Plasma Experiment on Pioneer 6 and Pioneer 7. M.I.T. Lab. for Nucl. Sci., Cambridge, Massachusetts, 1967.

The minimum detectable current at present is near or somewhat below 10^{-12} A; this limit is largely due to thermal noise in the circuits.

12.2.3.3. *Counting Methods.* More recent curved-plate analyzers do not measure currents but instead count individual particles leaving the exit aperture. The low energy (down to 100 eV and below) of the particles requires that they be detected without traversing any appreciable amount of matter; thus traditional counting methods such as Geiger tubes and scintillation counters are not usable (the thin-window Geiger tube extensively used in radiation belt observations, for instance, requires particles to penetrate 1.2 mg/cm^2 of matter, which limits it to electron energies above 40 keV). Particle detection is usually accomplished with some form of electron multiplier. In the multipliers used on the Vela satellites,[31] the particle hits a sensitive surface, producing secondary electrons which then cascade from electrode to electrode in the same manner as in the conventional photomultiplier. A more recent and simpler instrument of the multiplier type is the continuous channel multiplier (Bendix "channeltron").[32] The counting efficiency of all these instruments (ranging from approximately 10% to nearly 100% for particle energies usually encountered) must generally be determined by laboratory calibration; there also may be possible changes in efficiency with time and/or total charge collected, which may be particularly significant in the case of the channeltron.[33]

Another particle-counting system, developed by the Goddard–University of Maryland group,[11] postaccelerates the positive ions and lets them strike an aluminum target, producing secondary electrons (the postacceleration is achieved by placing a large negative potential on the suitably shaped target); the secondary electrons are then repelled by the negative potential of the target and strike a scintillator with enough energy to be counted. This system is intrinsically limited to the detection of positive particles and cannot be adapted to detect electrons.

The sole instance to date of counting methods used in conjunction with a Faraday cup is an instrument developed at Rice University.[34] A funnel-shaped continuous channel electron multiplier is the particle counter. The quantity measured is the number of counts when the modulator potential is high minus the number of counts when it is low; the value of this quantity is on the average proportional to the flux of particles with

---

[31] S. Singer, *Proc. IEEE* **53**, 1935 (1965).

[32] D. S. Evans, *Rev. Sci. Instrum.* **36**, 375 (1965).

[33] S. Cantarano, A. Egidi, R. Marconero, G. Pizzella, and F. Sperli, *Ric. Sci.* **37**, 387 (1967).

[34] J. W. Freeman, Jr., *J. Geophys. Res.* **73**, 4151 (1968).

energies per charge within the range defined by the upper and lower limits of the modulator potential, as in the conventional modulated Faraday cup. This method has the disadvantage that when most of the particles lie above the nominal energy window of the detector, the measured quantity is the small difference between two large counting rates and hence is subject to large statistical fluctuations. Such large fluctuations have actually been observed, and in fact used to detect the presence of large fluxes of particles above the energy range of the instrument.[35] (This problem does not arise with conventional modulated Faraday cups since the precision of the ac measurement is unaffected by the presence of direct currents due to unmodulated higher energy particles.)

12.2.4. Calibration

Before being placed on the spacecraft, plasma detectors are usually calibrated in the laboratory in order to determine the transmission function (or else to check a theoretically calculated one). Since all these instruments are simply detectors of particle beams rather than plasma detectors in the conventional sense, they can be calibrated by exposing them to particle beams obtained with standard beam technology; there is no need for the difficult task of producing an actual plasma, let alone one which would simulate solar wind conditions. Some special precautions should, however, be observed. The particle fluxes encountered by the detector are expected to be uniform over its entrance aperture; hence the calibration must be performed with effectively uniform illumination of the entrance aperture, obtained either directly with a wide, uniform beam or by scanning a narrow beam across the entire aperture. The spread of the beam in energy and angle must be small compared with the width of the transmission function; particularly with the complex and very narrow transmission functions of curved-plate analyzers, this requires beams with an energy spread $<1\%$ and an angular divergence $\lesssim 1°$.

12.2.5. Organization of Energy and Angle Measurements

The instruments described in the preceding sections can at any one time detect particles only within a limited range, centered about some mean value, of energy and angle. The mean energy can be varied by varying a suitable voltage in the instrument; usually the voltage is varied in preprogrammed steps (each step is often referred to as an "energy window" or "energy channel"). The angle can in principle be varied by

---

[35] J. W. Freeman, Jr., C. S. Warren, and J. J. Maguire, *J. Geophys. Res.* **73**, 5719 (1968).

varying the orientation of the detector and, to a more limited extent, by making use of various split-collector arrangements or by using several separate instruments looking in different directions. In designing a detector to be flown on a particular spacecraft, a set of energies and angles at which measurements are to be made must be selected and sequenced; this set must both be able to provide the desired information and be compatible with the limited capacity of the spacecraft telemetry. This section describes the principal schemes of organizing the complete set of energy and angle measurements. (For references to specific spacecraft, see Table II.)

12.2.5.1. Instruments on Spin-Stabilized Spacecraft. Many spacecraft are designed to spin stably about an axis fixed in inertial space. The typical spin period is of the order of a few seconds; for spacecraft intended to study the solar wind, the spin axis is usually oriented at a large angle to the ecliptic plane (the plane of the earth's orbit about the sun). A spinning spacecraft provides an easy method of varying the orientation of the detector, which is therefore mounted so that the normal to its entrance aperture forms a large angle (usually 90°) with the spin axis. The simplest scheme of taking measurements, and the one first used successfully, consists of maintaining a fixed energy window during one complete revolution of the spacecraft and measuring the flux as a function of angle (typically at 20 or so equally spaced angles); then the instrument is advanced to the next energy window and again during one revolution a complete angular scan is made, and so on until the instrument has cycled through all of its energy windows. This provides a complete two-dimensional set of measurements of flux as a function of energy and angle of azimuth about the spin axis (the dependence on polar angle can only be observed by such means as split collectors).

In the solar wind appreciable fluxes of protons are observed coming only from directions within a small range ($\sim 20°$) of the radius vector from the sun, and hence taking measurements at equally spaced angles is rather inefficient. Instruments designed primarily for study of the solar wind thus take many measurements at angles looking close to the sun and only a few (or else integrated over wide angular intervals) looking away from the sun. This scheme, of course, is possible only if the spacecraft is equipped with a sun sensor capable of indicating to the instrument when it is looking at the sun.

Another scheme, used with Faraday cups on Explorer 33 and 35 satellites, measures the total flux of protons (using a very wide energy window) as a function of angle, during one revolution, and also picks out the angle at which the maximum flux is coming; during succeeding revolutions it then measures the flux at that one angle as a function of energy. This provides two one-dimensional sets of measurements: total flux (integrated over energy) as a function of angle, and flux at one angle as a function of energy.

In all the schemes described so far, the time required to obtain a complete set of measurements (often called a complete "spectrum") is equal to the spin period times the number of energy channels, typically ~ 1 min. The interval between consecutive spectra could in principle be as short as the time required for one spectrum, but in practice it often is longer because of telemetry limitations.

A somewhat different scheme has to be used on spacecraft which rotate very slowly (such as Vela 4, which has a spin period of about 1 min). In this case the sequence of energy and angle measurements is reversed: the fluxes in all the energy windows are measured successively, as rapidly as possible and hence all at essentially one angle; this is then repeated at other angles, and a complete energy-angle scan is obtained during one (in this case rather long) revolution of the spacecraft.

A convenient way of labeling a single measurement at a particular energy and angle is to specify the mean energy per charge E/q accepted by the particular energy window and the angles (θ, ϕ) which define the normal direction of the entrance aperture with respect to some fixed (i.e., not rotating with the spacecraft) coordinate system; one may then define a corresponding speed u by the equation

$$e|E/q| = \tfrac{1}{2}mu^2,$$

where e is the elementary charge and m the mass of the proton (or the electron if negative particles are being measured), and consider the triplet of numbers (u, θ, ϕ) as a vector $\mathbf{u}$ of magnitude u and direction given by (θ, ϕ). (In principle a third angle ψ, the angle of rotation of the detector about its normal direction, should also be specified if the detector's transmission function is not axially symmetric, but in nearly all cases occurring in practice the motion of the detector is constrained so that ψ is a unique function of θ and ϕ.) Then each measurement corresponds to a single "velocity" vector $\mathbf{u}$ and can be thought of as a point in $\mathbf{u}$ space. The various sampling schemes discussed above can then be succinctly represented by series of points in the plane in $\mathbf{u}$ space perpendicular to the spacecraft spin axis. These representations are illustrated in Fig. 8.

12.2.5.2. *Instruments on Triaxially Stabilized Spacecraft.* Some spacecraft (notably the Mariner series and the later satellites in the OGO series) do not spin but maintain a fixed orientation in space, with only very slow,

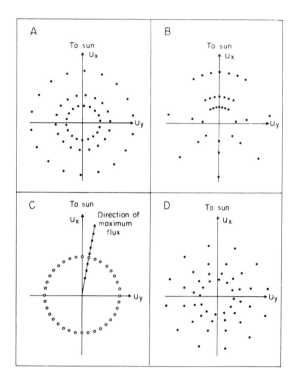

FIG. 8. Sampling schemes for energy and angle measurements on rotating spacecraft. (A) Measurements made at equally spaced angles for each energy window. (B) Measurements concentrated at angles close to the solar direction. (C) The Explorer 33–35 scheme: open circles represent measurements in very wide ("integral") energy windows; dots represent measurements in narrow ("differential") energy windows. (D) Measurements on a slowly rotating spacecraft.

long-term changes. Hence if measurements of flux as a function of angle are desired, they must be made by means of split-collector or multiple-detector arrangements (scanning by means of mechanical motion of the detector is, in principle, also possible but to date has not been successfully realized), and the amount of angular information available from these spacecraft is generally quite inferior to that from spin-stabilized spacecraft. On the other hand, a much higher time resolution in the measurement of flux as a function of energy at a fixed angle is possible since one is not constrained to wait a full rotation before advancing the instrument from one energy step to the next. (Alternatively, a much longer integration time for a single measurement is possible.)

12.3. Methods of Analysis

12.3.1. Nature of Information To Be Obtained

In this section we discuss how and what information about the plasma can be obtained from measurements of currents or counting rates at various energies and angles. Referring back to Table I, we note that all extraterrestrial plasmas accessible to direct measurement are collisionless, in the sense that the mean free path for Coulomb collisions is much longer than the characteristic macroscopic length scales. Hence there is no *a priori* reason to expect any kind of thermodynamic equilibrium or applicability of the temperature concept; a complete description of the plasma requires the specification of the velocity distribution function for each species of particles present. Nevertheless, it has been found empirically that, at least in the solar wind and the magnetosheath, the plasma can be described to a surprisingly adequate extent for many purposes by specifying only its hydrodynamic parameters: density, bulk velocity, pressure tensor, and (sometimes) heat flux vector.[1,14] (These may have to be specified separately for each species of particles; physical considerations require that the number densities and bulk velocities of positive and negative particles to a very high approximation be equal, but the pressures and heat fluxes for individual species may very well be different and have in fact been observed to be different.[1,7]) Much of the analysis of extraterrestrial plasma measurements consists of determining these hydrodynamic parameters of the plasma, which of course are simply the moments of the velocity distribution function $f(\mathbf{v})$: if n is the number density, $\mathbf{V}$ the bulk velocity, $\mathbf{P}$ the pressure, and $\mathbf{q}$ the heat flux, then these are *defined* for each species by

$$n = \int d\mathbf{v}\, f(\mathbf{v}),$$

$$n\mathbf{V} = \int d\mathbf{v}\, \mathbf{v} f(\mathbf{v}),$$

$$\mathbf{P} = m \int d\mathbf{v}\, f(\mathbf{v})(\mathbf{v} - \mathbf{V})(\mathbf{v} - \mathbf{V})$$

$$\mathbf{q} = \tfrac{1}{2} m \int d\mathbf{v}\, f(\mathbf{v})(\mathbf{v} - \mathbf{V}) |\mathbf{v} - \mathbf{V}|^2$$

(12.3.1)

(where the integrals are volume integrals over velocity space). All extraterrestrial plasmas contain a magnetic field and the pressure tensor is usually assumed to be axially symmetric:

$$\mathbf{P} = P_\perp \mathbf{I} + (P_\parallel - P_\perp)\mathbf{bb}$$

where **I** is the unit dyadic and **b** is the unit vector in the direction of the field. The "temperature" is simply defined, for convenience, as the ratio $kT \equiv \mathbf{P}/mn$ and is thus also a tensor; the use of the term "temperature" does not in any way imply the existence of thermal equilibrium or a Maxwellian velocity distribution. (The reader, however, should beware of some observational papers in which the word "temperature" is applied to various specialized measures of velocity dispersion having only a vague connection with the term as used in plasma physics.)

Besides the moments of the distribution function, its detailed structure is also of interest but can generally be studied only in regions of space, such as the magnetosheath and especially the outer magnetosphere, where the distribution function is very broad in comparison with the transmission functions of available detectors.

12.3.2. Relation of Measurements to the Particle Distribution Function

The raw data obtained from a plasma detector on a spacecraft consists of a sequence of currents or counting rates at various energy windows and angular orientations of the detector. It is convenient to specify the energy and angle of each measurement by the vector **u** described in Section 12.2.5.1. If $C(\mathbf{u})$ is the counting rate of the detector during a particular measurement, then the contribution to it of particles of a given velocity is equal to the number of particles at that velocity, times the component of velocity normal to the entrance aperture, times the area of the entrance aperture, times the transmission function. (We are now assuming that, as discussed in Section 12.1.2, the flux on the entrance aperture represents the plasma particle flux unaffected by the spacecraft and hence is uniform over the aperture, which is much smaller than any length scale of the plasma.) To obtain the total counting rate it is necessary to sum the contributions from all velocities and all species of particles (unless the instrument is designed to accept only one species), which means an integral over velocity space weighted by the sum of velocity distribution functions for all species. Thus (assuming that the counting efficiency for particles of each species is the same), we obtain the equation

$$C(\mathbf{u}) = S \sum_a \int d\mathbf{v} \, v_n G(\mathbf{u}(Z_a/A_a)^{1/2}, \mathbf{v}) f_a(\mathbf{v}) \qquad (12.3.2)$$

where Z_a and A_a are the atomic mass numbers of the ath species, S is the area of the entrance aperture, $v_n \equiv \mathbf{v} \cdot \mathbf{u}/u$ is the velocity component normal to the entrance aperture, $f_a(\mathbf{v})$ is the velocity distribution function of particles of the ath species, and $G(\mathbf{u}, \mathbf{v})$ is related to the detector transmission function as follows:

12.3. METHODS OF ANALYSIS

The transmission function is specified as a function of proton (or electron) speed, direction of incidence relative to the detector, and the quantity u giving the energy per charge window (see Section 12.2.5.1); it can thus be written as the function $H(\mathbf{v'}, u)$, where $\mathbf{v'}$ is the particle velocity vector expressed in a coordinate system Σ' fixed with respect to the detector. If $\mathbf{v}$ is the velocity vector expressed in some suitable coordinate system Σ fixed in space (i.e., not rotating with the spacecraft) and the detector orientation with respect to Σ is given by the direction of $\mathbf{u}$, then

$$G(\mathbf{u}, \mathbf{v}) \equiv H(\mathbf{R} \cdot \mathbf{v}, u) \qquad (12.3.3)$$

where $\mathbf{R}$ is the rotation matrix that transforms a vector from Σ to Σ'; clearly $\mathbf{R}$ depends on the angles θ, ϕ specifying the direction of $\mathbf{u}$, and on the angle ψ if relevant (see Section 12.2.5.1). If the width of the transmission function in energy is proportional to the energy, i.e., the quantity $\Delta E/E$ discussed in Section 12.2.2 is independent of E (this is the case for most curved-plate analyzers and for Faraday cups in which the ratio of ac to dc components of the modulating voltage is constant), then $G(\mathbf{u}, \mathbf{v})$ depends only on the directions of $\mathbf{u}$, $\mathbf{v}$ and on the ratio of the magnitudes u/v; thus it can be written as

$$G(\mathbf{u}, \mathbf{v}) = G^*(\mathbf{u}/u, \mathbf{v}/u) \qquad (12.3.4)$$

If the detector does not count individual particles but measures currents, the contribution of each species is weighted by its charge and Eq. (12.3.2) is replaced by the following equation for the current $I(\mathbf{u})$:

$$I(\mathbf{u}) = eS \sum_a Z_a \int d\mathbf{v}\, v_n G(\mathbf{u}(Z_a/A_a)^{1/2}, \mathbf{v}) f_a(\mathbf{v}) \qquad (12.3.5)$$

The velocity distribution function is also a function of space and time; hence, if the measurement is made at a time t while the spacecraft is located at position $\mathbf{r}$, strictly speaking $f(\mathbf{v})$ in the above equations should be written as:

$$(1/\tau) \int_t^{t+\tau} dt'\, f(\mathbf{r}, \mathbf{v}, t'), \qquad (12.3.3)$$

where τ is the duration of the measurement (the change in $\mathbf{r}$ during an interval τ is, in practice, negligible). For the type of detailed analysis described later in this section to be meaningful, it is necessary that the distribution function not change significantly during the time required for the acquisition of one complete set of measurements; as described earlier, for present instruments this time is of the order of minutes. If the plasma changes appreciably on shorter time scales, the time and velocity dependences of the distribution function cannot be distinguished. (If, however, the changes are all random fluctuations on a time scale much shorter than the duration τ of a single measurement, e.g., high frequency plasma waves, then the analysis is again meaningful but it now yields information only on the distribution function averaged over the fluctuations.)

The contributions to the measurements of different ion species can be separated in principle only by using a suitable instrument, such as the velocity selector. In practice, the predominant positive component of extraterrestrial plasma is protons with a small admixture of alpha particles (approximately 4–10% by number in the solar wind).[1] As can be seen from Eqs. (12.3.2) and (12.3.5), alpha particles are detected at voltages which are twice those for protons, for the same velocity distribution functions. In the solar wind both distribution functions are narrow and the proton and alpha contributions appear as two well-separated peaks in the energy per charge spectrum; thus the two species can be distinguished even without special instruments. To simplify the following discussion, we shall treat the plasma as consisting only of protons and electrons; extensions to other species are straightforward.

Instead of the current $I(\mathbf{u})$ or counting rate $C(\mathbf{u})$ it is convenient to deal with an equivalent quantity having the dimensions of a flux density, $F(\mathbf{u})$, defined as

$$F(\mathbf{u}) \equiv C(\mathbf{u})/S \quad \text{or} \quad F(\mathbf{u}) \equiv I(\mathbf{u})/eS,$$

which is sometimes loosely referred to as "the measured flux." Equations (12.3.2) and (12.3.5) then both become (considering only one species)

$$F(\mathbf{u}) = \int d\mathbf{v}\, v_n G(\mathbf{u}, \mathbf{v}) f(\mathbf{v}). \tag{12.3.6}$$

The basic unfolding problem of extraterrestrial plasma measurements then is: given the measured values $F(\mathbf{u})$ at a discrete set of $\mathbf{u}$, what information can be obtained about the distribution function $f(\mathbf{v})$?

If $F(\mathbf{u})$ were known for all values of $\mathbf{u}$ (considered as a continuous variable), then Eq. (12.3.6) would be a three-dimensional integral equation for $f(\mathbf{v})$; it can be shown, for the special case of a detector whose response is axially symmetric and of the form given by Eq. (12.3.4), that the integral equation (12.3.6) can be explicitly inverted by expanding all functions in spherical harmonics and taking Mellin transforms in energy. This explicit inversion, however, appears to be of no practical use whatever, and approximate methods must be used to solve the unfolding problem.

12.3.2.1. Differential and Integral Behavior. There are two limiting cases in which Eq. (12.3.6) assumes a much simpler form. The first occurs when the distribution function has a much greater width in velocity space than the transmission function and $f(\mathbf{v})$ in (12.3.6) does not vary appreciably over the range of **v** for which $G(\mathbf{u}, \mathbf{v})$ is different from zero; then $f(\mathbf{v}) \approx f(\mathbf{u})$ can be taken out of the integral and (12.3.6) rewritten as

$$F(\mathbf{u}) = [f(\mathbf{u})u^2/m]mu^2\xi,\tag{12.3.7}$$

$$\xi \equiv u^{-4} \int d\mathbf{v}\, v_n G(\mathbf{u}, \mathbf{v}).$$

The quantity in brackets will be recognized as the flux density per unit energy of particles with velocity **u**, i.e., the differential directional intensity familiar in radiation belt studies. ξ is dimensionless and, if G can be written in the form (12.3.4), depends only on the direction and not the magnitude of **u**; $u^3\xi$ can be thought of as the effective volume of the transmission function in velocity space.* (Note, however, that $u^3\xi$ in general *cannot* be written in the at first sight "obvious" manner as $\Delta v_n \Delta v_{t1} \Delta v_{t2}$, where Δv_n and Δv_{t1}, Δv_{t2} are the ranges of velocity along the normal and the two transverse directions accepted by the detector, or in the equivalent polar coordinate form $u^2\, \Delta v_n\, \Delta\theta\, \Delta\phi$, since the range of acceptance in one velocity component may strongly depend on the other components, as illustrated in Fig. 7.) Thus in this limiting case the "measured flux" $F(\mathbf{u})$ is proportional to the differential flux at the point in velocity space corresponding to the center of the detector energy window and angular orientation, with a constant of proportionality closely related to the width of the transmission function. Measurement of $F(\mathbf{u})$ as a function of **u** thus directly yields the distribution function.

The other limiting case occurs in the opposite extreme, when the distribution function is very narrow in comparison with the transmission function; this ordinarily requires that the spread of particle velocities about the bulk velocity of the plasma be very small compared to the bulk velocity itself. In this case an appreciable measured flux is obtained only if the plasma bulk velocity **V** lies within the detector energy-angle window, and then $G(\mathbf{u}, \mathbf{v})$ may be assumed not to vary significantly over the range of **v** for which $f(\mathbf{v})$ is appreciably different from zero, so that $G(\mathbf{u}, \mathbf{v}) \approx G(\mathbf{u}, \mathbf{V})$ may be taken out of the integral in (12.3.6):

$$F(\mathbf{u}) \approx G(\mathbf{u}, \mathbf{V}) \int d\mathbf{v}\, v_n f(\mathbf{v}) = G(\mathbf{u}, \mathbf{V}) n V_n \tag{12.3.8}$$

where n is the number density of the particle species being observed and V_n is the component of **V** normal to the entrance aperture [cf. Eq. (12.3.1)]. Thus the "measured flux" $F(\mathbf{u})$ is now proportional to the total flux of

---

[36] G. C. Theodoridis and F. R. Paolini, *Rev. Sci. Instrum.* **40**, 621 (1969).

---

* For an extensive description of ξ for curved-plate analyzers, see Theodoridis and Paolini[36] and references therein.

particles; the constant of proportionality is independent of the *width* of the transmission function and depends only on its value at the energy and direction corresponding to the bulk velocity of the plasma. Measurement of $F(\mathbf{u})$ as a function of $\mathbf{u}$ thus serves to fix $\mathbf{V}$ (most easily obtained as the value of $\mathbf{u}$ for which the largest value of F occurs) but provides no information on the shape of the distribution function; the shape of $F(\mathbf{u})$ simply reflects the transmission function.

The first limiting case, in which the detector provides differential measurements of the flux in velocity space, is generally encountered in the outer magnetosphere. The second limiting case, in which the detector provides a measurement of the total integrated flux, is a first approximation to the behavior of all detectors (except those with extremely narrow transmission functions) in the solar wind. In the magnetosheath generally (and for some detectors in the solar wind), the distribution function and the transmission function have comparable widths and the full integral in Eq. (12.3.6) must be retained.

It should also be noted that many detectors have transmission functions which are narrow in one direction and wide in another. For example, hemispherical curved-plate analyzers generally have a very narrow response in energy (approximately a few percent) and azimuthal angle (approximately a few degrees) but a very wide response in polar angle (nearly 90°); a modulated Faraday cup may select particles in a very narrow range of velocity normal to its aperture but accept a wide range of transverse velocities. A more complex case occurs with the quadrispherical analyzer, whose transmission function has a nonzero value over a region of velocity space that is greatly elongated along a direction approximately 70° from normal incidence (see Fig. 7). If the width of the transmission function along its "wide" and along its "narrow" direction is much larger and much smaller, respectively, than the width of the distribution function, the "measured flux" $F(\mathbf{u})$ is proportional to the flux integrated over the "wide" direction but differential along the "narrow" direction. As a function of $\mathbf{u}$, $F(\mathbf{u})$ then has a hybrid character: its dependence on the component of $\mathbf{u}$ along a "narrow" direction reflects the behavior of $f(\mathbf{v})$, while its dependence on the component of $\mathbf{u}$ along a "wide" direction simply reproduces the transmission function. If the existence of these different widths in the transmission function is not appreciated, the resulting behavior of $F(\mathbf{u})$ may be falsely attributed to an anisotropy in the distribution function.*

[37] A. J. Hundhausen, *J. Geophys. Res.* **74**, 3740 (1969).

* Compare the discussion of Pioneer 6 measurements by Hundhausen.[37]

12.3.3. Estimation of Plasma Parameters

12.3.3.1. Method of Moments. As discussed above, most of the effort in data analysis, particularly in the case of the solar wind, has gone into estimating the hydrodynamic parameters of the plasma (moments of the distribution function). The simplest method for estimating these is obtained if one assumes that the detector can, to some approximation, be treated as differential; then, if the measurements are made at energies and angles specified by $\mathbf{u}_j$ ($j = 1, 2, ..., N$), each measurement provides a value of the distribution function $f(\mathbf{v})$ according to Eq. (12.3.7). Now a particular moment M_p of $f(\mathbf{v})$ can be approximated by a sum

$$M_p \equiv \int d\mathbf{v}\, \mathbf{v}^p f(\mathbf{v}) \approx \sum_{j=1}^{N} C_j \mathbf{u}_j^p f(\mathbf{u}_j),$$

where the C_j are constants obtained from the usual theory of approximate numerical quadrature; substituting for $f(\mathbf{u}_j)$ from Eq. (12.3.7), we obtain an estimate of the moment M_p as a linear combination of the measured fluxes:

$$M_p \approx \sum_{j=1}^{N} \alpha_i^{(p)} F(\mathbf{u}_j),$$

$$\alpha_j^{(p)} \equiv C_j \mathbf{u}_j^p / u_j^4 \xi.$$

(12.3.9)

This method has been extensively used to analyze measurements in the outer magnetosphere, where the assumption of differential detector behavior is usually well justified. It has also been used with some success to obtain the proton density and bulk velocity (but not pressure) from measurements in the solar wind.[38,39] As the assumption of differential behavior is in this case clearly inappropriate, the justification of the method is purely empirical and the coefficients $\alpha_j^{(p)}$ in Eq. (12.3.9) are obtained empirically by requiring the method to yield the correct results when applied to a set of $F(\mathbf{u}_i)$ calculated (using Eq. (12.3.8)) from a model distribution of known density and bulk velocity; in this case the extreme computational speed of the method of moments is its sole advantage over the method described in the following section.

12.3.3.2. Use of Model Distributions. By far the most extensively used method has been to assume a specific functional form of the distribution function containing some adjustable parameters, calculate the expected

---

[38] E. F. Lyon, A study of the interplanetary plasma. Ph.D. Thesis, M.I.T., Cambridge, Massachusetts, 1966 (unpublished).

[39] A. J. Lazarus, H. S. Bridge, J. M. Davis, and C. W. Snyder, *Space Res.* **7**, 1296 (1967).

fluxes from Eq. (12.3.8) as functions of the parameters, and then vary the parameters until the predicted and the measured fluxes agree as closely as possible. (The adjustable parameters very often are the moments of the distribution themselves.) While this method is not expected to yield precise information on the detailed shape of $f(\mathbf{v})$, the estimated moments of the distribution function should be relatively accurate provided that its actual shape does not greatly differ from the assumed model and that a significant fraction of the particles lies within the region of velocity space surveyed by the detector. For the use of this method to be meaningful it is, of course, necessary that there be more measurements than adjustable parameters to be fitted to them.

Various methods of fitting have been used. The classical least-squares technique is the most straightforward, although rather cumbersome and time-consuming. Simplified methods, generally tailored to the specific characteristics of a particular set of observations, have been evolved[15,17,40]; most of them estimate the parameters from a few selected measurements and use the remaining measurements to verify and/or improve these estimates.

Among the principal model distributions used to date are the following:

(1) Convected Maxwellian:

$$f(\mathbf{v}) = \frac{n}{\pi^{3/2} w_0^3} \exp\left\{ -\frac{(\mathbf{v} - \mathbf{V})^2}{w_0^2} \right\}.$$

Adjustable parameters are: density n, bulk velocity $\mathbf{V}$, and most probable speed w_0 (related to temperature T by $kT = \tfrac{1}{2} m w_0^2$). This is the simplest model and the first one used.[41] When used to fit solar wind data from detectors whose transmission functions are much wider than the proton distribution function, it yields good values of the density and bulk velocity (more accurate than those obtained by the method of moments) plus a rough estimate of the width of the distribution function.

(2) Convected bi-Maxwellian[24,42,43]:

$$f(\mathbf{v}) = \frac{n}{\pi^{3/2} w_{0\|} w_{0\perp}^2} \exp\left\{ -\frac{w_\|^2}{w_{0\|}^2} - \frac{w_\perp^2}{w_{0\perp}^2} \right\},$$

$$\mathbf{w} \equiv \mathbf{v} - \mathbf{V}, \qquad w_\| = \mathbf{w} \cdot \mathbf{b}, \qquad \mathbf{w}_\perp = \mathbf{w} - \mathbf{b} w_\|.$$

[40] M. Neugebauer and C. W. Snyder, *J. Geophys. Res.* **71**, 4469 (1966).
[41] F. Scherb, *Space Res.* **4**, 797 (1964).
[42] F. L. Scarf, J. H. Wolfe, and R. W. Silva, *J. Geophys. Res.* **72**, 993 (1967).
[43] V. Formisano, *J. Geophys. Res.* **74**, 355 (1969).

Adjustable parameters are: density n, bulk velocity $\mathbf{V}$, and parallel and perpendicular thermal speeds $w_{0\|}$ and $w_{0\perp}$; $\mathbf{b}$ is the unit vector along the magnetic field and is generally known from simultaneous measurements on the same spacecraft. This appears to be, at present, a reasonably close approximation to the actual proton distribution function in the solar wind.

(3) The function

$$f(\mathbf{v}) = \frac{n}{\pi^{3/2}w_0^3} \frac{\Gamma(\kappa + 1)}{\kappa^{3/2}\Gamma(\kappa - \tfrac{1}{2})} \left[1 + \frac{|\mathbf{v} - \mathbf{V}|^2}{\kappa w_0^2}\right]^{-(\kappa + 1)}$$

which is a generalization of the "resonance" distributions used in plasma physics (the case $\kappa = 2$ is also known as the Cauchy distribution) and is sometimes known in the trade as the "κ-distribution." The adjustable parameters are the density n, the bulk velocity $\mathbf{V}$, the most probable speed w_0, and the exponent κ; at large values of v ($|\mathbf{v} - \mathbf{V}|^2 \gg w_0^2$) the distribution function varies approximately as (energy)$^{-\kappa - \frac{1}{2}}$ [the corresponding differential intensity varies as (energy)$^{-\kappa}$]. As $\kappa \to \infty$, $f(\mathbf{v})$ approaches the Maxwellian distribution. This function has been used mainly to analyze measurements of electrons in the magnetosheath and the outer magnetosphere (in the latter case the bulk velocity is usually assumed *a priori* to be negligible), in order to represent the effects of a large non-Maxwellian high-energy "tail" that appears to be present in the electron distribution function within these regions.[14,17,44]

As an example of the use of this method, Fig. 9 shows two sets of Faraday cup measurements on the IMP-1 satellite, together with theoretically calculated fluxes from the fitted models.

12.3.3.3. Construction of the Distribution Function by Interpolation. As already pointed out in Section 12.3.2, if the distribution function is very wide compared to the transmission function, the measured flux $F(\mathbf{u})$ directly gives the distribution function. If the widths of the transmission function and the distribution function are comparable, it may still be possible to bypass model construction and obtain the distribution function at a particular point by interpolating between several measured fluxes. Techniques for doing this have been developed by Hundhausen *et al.*[45] and by Ogilvie *et al.*[46] In essence, one considers a few adjacent measure-

---

[44] A. J. Lazarus, G. L. Siscoe, and N. F. Ness, *J. Geophys. Res.* **73**, 2399 (1968).

[45] A. J. Hundhausen, J. R. Asbridge, S. J. Bame, H. E. Gilbert, and I. B. Strong, *J. Geophys. Res.* **72**, 87 (1967).

[46] K. W. Ogilvie, L. F. Burlaga, and H. Richardson, Goddard Space Flight Center Rep. X-612-67-543. Goddard Space Flight Center, Greenbelt, Maryland, 1967.

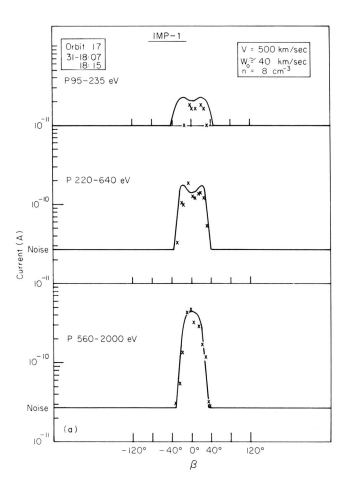

FIG. 9. A set of Faraday cup measurements on IMP-1 and model distributions fitted to them. Shown are currents measured in each energy window as a function of spacecraft rotation angle β; + and × are measured values, lines are theoretical fits. Above: Proton measurements in the solar wind. A convected Maxwellian has been assumed. Right: Measurements in the magnetosheath. The "contamination" of proton measurements by electrons (see Section 12.2.2.1) is in this case significant. A convected Maxwellian has been assumed for protons and a κ-distribution for electrons. Dashed lines: predicted currents due to protons alone. Solid lines: predicted currents due to protons and electrons [from S. Olbert, in "Physics of the Magnetosphere" (R. L. Carovillano, J. F. McClay, and H. R. Radoski, eds.), p. 641. Reidel, Dordrecht, Holland, 1969].

12.3. METHODS OF ANALYSIS

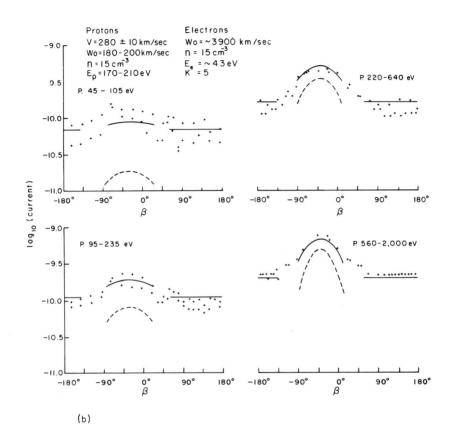

(b)

FIG. 9.

ments at a time [for example $F(\mathbf{u})$ for $\mathbf{u}_{j-1}$, $\mathbf{u}_j$, $\mathbf{u}_{j+1}$] and assumes that both the function $F(\mathbf{u})$ for $\mathbf{u}$ near $\mathbf{u}_j$ and the distribution function $f(\mathbf{v})$ for some narrow range of $\mathbf{v}$ about $\mathbf{v} = \mathbf{u}_j$ can be represented by Gaussian functions (the transmission function is also usually approximated by a Gaussian or by a step function); then it is possible to solve for the value of $f(\mathbf{v})$ at $\mathbf{v} = \mathbf{u}_j$ in terms of the adjacent measured values $F(\mathbf{u})$ for $\mathbf{u} = \mathbf{u}_{j-1}$, $\mathbf{u}_j$, $\mathbf{u}_{j+1}$. By taking a series of adjacent groups of measurements, values of $f(\mathbf{v})$ at a series of $\mathbf{v} = \mathbf{u}_j$ for all j can be obtained; values of $f(\mathbf{v})$ at intermediate values of $\mathbf{v}$ can then be obtained by conventional interpolation, and moments of $f(\mathbf{v})$ can be calculated, if desired, by numerical integration over the function thus obtained. An example of a solar wind proton distribution function constructed in this manner from curved-plate analyzer measurements on the Vela 3 satellite is shown in Fig. 10.*

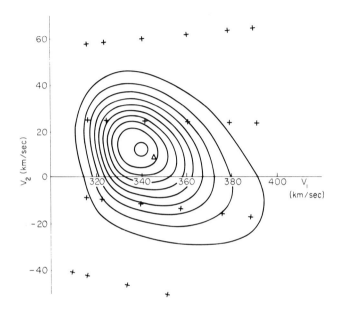

FIG. 10. Contour map of a proton velocity distribution function in the solar wind obtained from a set of Vela 3 curved-plate analyzer measurements. The function is normalized to a maximum value of 1 and the contours are drawn at intervals of $\frac{1}{10}$. The triangle shows the bulk velocity. The ×'s indicate the the points $\mathbf{u}_j$ at which the actual measurements were made; the width of the transmission function along the V_1 or V_2 direction is comparable to the spacing of the +'s along that direction [from A. J. Hundhausen, J. R. Asbridge, S. J. Bame, H. E. Gilbert, and I. B. Strong, *J. Geophys. Res.* **72**, 87 (1967)].

---

* ACKNOWLEDGMENTS: I am grateful to Professors H. S. Bridge, S. Olbert, and A. J. Lazarus for critical reading of the manuscript. This work was supported by the National Aeronautics and Space Administration under grant NGL 22–009–015.

13. WHISTLERS: DIAGNOSTIC TOOLS IN SPACE PLASMA†

13.1. Introduction

In 1919 Barkhausen,[1] using equipment that consisted, in essence, of a pair of earphones and an amplifier connected to a long wire, heard peculiar musical tones with continuously descending pitch. This phenomenon remained largely a scientific curiosity until 1953, when Storey[2] suggested that these "whistlers" were electromagnetic waves radiated by lightning discharges and subsequently dispersed by propagation in the earth's ionized outer atmosphere. Storey deduced that the waves followed magnetic field lines from one hemisphere of the earth to the other and showed that, at the lowest frequencies, the whistler propagation delay, t, is given by

$$t = Kf^{-1/2} \int_{\text{path}} (N/B)^{1/2} \, ds, \qquad (13.1.1)$$

where f is the wave frequency, N the electron density, B the magnetic field strength, ds an element of the propagation path, and K is a known constant.

Storey noted that whistlers were uniquely suited to measuring electron densities at large distances in the earth's outer atmosphere, within the magnetosphere. The propagation delay is proportional to the square root of the density, weighted inversely as the square root of the magnetic field strength. Thus the high electron density occurring over a relatively short distance in the ionosphere where the magnetic field is large makes only a minor contribution to the propagation time. At high altitudes, the path is long and the magnetic field weak so that most of the propagation delay occurs in this region.[3]

An additional (extremely) useful happenstance is that the plasma flow is largely constrained to be in the same direction as the magnetic field, which is also the propagation path of the whistler. Thus the whistler technique is uniquely suited to measuring the total ionization content of a

---

[1] H. Barkhausen, *Phys. Z.* **20**, 401 (1919).
[2] L. R. O. Storey, *Phil. Trans. Roy. Soc. London, Ser. A.* **246**, 113 (1953).
[3] D. L. Carpenter and R. L. Smith, *Rev. Geophys.* **2**, 415 (1964).

---

† Part 13 by Neil M. Brice and Robert L. Smith.

magnetic flux tube in the upper atmosphere. This property also simplifies the problem of obtaining models of the electron density distribution along the propagation path, and from these and the (known) magnetic field strength, the contributions to the total propagation delay of the various segments of the path.

In recent years, the whistler technique has been expanded to provide information on a wide variety of parameters relating to this region of space. These include not only the magnitude of the electron density and its spatial and temporal variation (including irregularities and gradients in density) but also ion density, composition, and temperature. Movement of ionization can also be detected and interpreted as a measure of large scale quasi-static electric fields. The breadth of information obtained arises in part from the nature of the radio waves emanating from a lightning discharge. The duration of the discharge (of the order of 10 msec) is typically much less than the duration of the whistler (of the order of 1 sec). Thus analysis of the radiation received at some distance from the source provides information on the "impulse response" of the intervening medium. This in turn is interpreted in terms of the parameters of the propagation medium.

We approach this subject from a "plasma diagnostic tool" viewpoint, with the plasma of interest being the earth's ionosphere and magnetosphere. In this perspective, the measurements of whistler propagation delay are quite accurate—of the order of 1 or 2%, with agreement between theory and observation to about the same order of accuracy. Electron densities in the magnetic equatorial plane are determined by ground-recorded whistlers with uncertainties of about $\pm 30\%$ and flux tube content to about $\pm 20\%$, while ion whistlers recorded in satellites may be used to measure the magnetic field strength to an accuracy of $\pm 0.1\%$, and ion density and temperature with uncertainties of about $\pm 20\%$ and $\pm 30\%$, respectively. Since both the experimental method and relevant theory are relatively simple by comparison with other areas of plasma physics (perhaps deceptively so), we will concentrate our attention on data analysis. Following a small section on the experimental method, we examine different types of whistlers phenomenologically, with the phenomena being organized primarily according to the information they provide.

13.2. Experimental Method

The object is to record naturally occurring electromagnetic waves in the frequency range from about 30 Hz to 30 kHz. The sensors used range from multiturn loop antennas as small as 1 ft in diameter fitted

13.2. EXPERIMENTAL METHOD

into rocket nose cones to a dipole several miles long laid on the antarctic ice. Typical antenna dimensions for either ground-based or satellite-borne loops or dipoles are one to ten meters.

These sensors are much smaller than the wavelength of the whistler waves so that the loops are sensitive essentially only to the magnetic field of the waves, while the dipoles are sensitive only to the electric field. The sensors are connected to low-noise preamplifiers with a very low impedance input for loops and very high impedance for dipoles.

The minimum detectable power of the whistler signal, which is determined by antenna size and receiver noise, is typically about 3×10^{-13} watt/m^2 for a 30 kHz receiver bandwidth. The loudest whistlers are about one million times stronger. For ground-based recordings, the strongest signals (impulsive atmospherics from nearby lightning discharges) may be a million times stronger yet. Therefore careful attention must be paid to the dynamic range and overload characteristics of the recording system.

The data are usually recorded directly (broad-band) on magnetic tape, together with calibration and timing information. Signals from low frequency transmitters also serve as time references, especially for the comparison of recordings made in separate places.

The tapes may be monitored aurally or replayed into a spectrum analyzer which produces an intensity modulated frequency-time spectrogram. Dynamic spectra of whistlers are shown in Figs. 2, 9, 16, 18, 19, 20, 26, 27, and 28. Spectrograms such as these form the basis for most whistler data analysis. The whistler technique has a number of inherently useful properties. As no transmitter is used, power requirements are minimal. Propagation delays are of the order of a second or more, and are therefore easily and accurately measured. Also, frequencies of interest are in the audio-frequency range for which a high-quality technology is available at relatively low cost. A typical broadband-recording whistler installation would include an antenna, preamplifier, main electronics and tape recorder, together with associated timing and control equipment. The cost of the initial installation (capital equipment and spare parts) is about $10,000 for a basic system with annual operating expenses about $3000 for magnetic tape and $1500 for replacement spare parts and small on-site equipment improvements.

For ground-based whistler observatories, site selection plays an important role. Whistler rates are highest at mid- to high-latitudes, in regions magnetically conjugate to areas of high-thunderstorm activity. Also, interference at harmonics of power-line frequencies usually requires that the site be located well away from heavily populated areas. Interference from other on-board systems is also a major problem for satellite- and rocket-borne receivers.

13.3. Ground-Based Whistler Observations

13.3.1. Whistler Echo Trains

Figure 1 illustrates the propagation of whistlers recorded by ground-based observatories. A lightning discharge occurs in, say, the Northern Hemisphere. Part of the radiated energy propagates between the earth and ionosphere and arrives as an impulse at time $t = 0$ at station A. This impulse (usually called the causative atmospheric) appears as a vertical

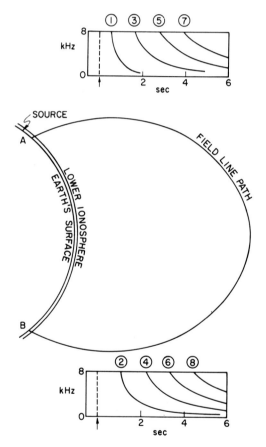

Fig. 1. Diagram of a whistler echo train. Energy from the impulsive lightning source in the Northern Hemisphere propagates partly in the earth–ionosphere wave-guide and is received as a strong impulse (dashed vertical line over arrow in upper inset) at station A and a weak impulse (dashed vertical line in lower inset) at station B. Part of the energy propagates in the magnetosphere along a magnetic field line where the energy may be reflected from hemisphere to hemisphere.

line on a frequency-time spectrogram shown idealized in the upper part of Fig. 1. A small fraction of the energy propagates in the earth-ionosphere wave-guide to the opposite hemisphere where it is recorded at station B as a very weak slightly delayed impulse. This is illustrated in the lower part of Fig. 1. Part of the energy penetrates the ionosphere in the vicinity of the discharge and propagates in the whistler mode along a magnetic field-line path to the opposite hemisphere. Here part of the energy leaks out of the ionosphere and is recorded at station B.

During the propagation in the ionosphere and magnetosphere the velocity of propagation is a strong function of frequency and generally much less than the velocity of light. For frequencies much less than the minimum electron cyclotron frequency the dispersed impulse or whistler is recorded as a descending gliding tone. This whistler is called a "one-hop" whistler and the corresponding first trace on the lower spectrogram in Fig. 1 is designated accordingly. Some of the energy is reflected back along the magnetic field-line path to produce a "two-hop" whistler at station A. Further reflections may lead to whistler echo trains.

The dispersions of whistlers in the echo trains are in the ratio 1:3:5, etc., if the receiver is in the opposite hemisphere from the source, or 2:4:6, etc., for a source in the same hemisphere. In observed whistler trains the measured dispersions are, within experimental error, in precise agreement with the integral ratios above. This implies that on each successive traversal of the outer ionosphere the wave energy follows the same path, i.e., is constrained to flow within a single (presumably field-aligned) "duct." The physical nature and properties of ducts are described below.

13.3.2. Nose Whistlers

Frequently energy from a single lightning discharge will propagate in several different ducts, producing a whistler with several discrete components. Figure 2 shows two spectrograms of particularly fine examples of multi-component whistlers recorded at Eights Station in Antarctica. Many of the components show both a rising and falling tone which are contiguous at a frequency of minimum propagation delay. Whistlers showing this characteristic are called "nose" whistlers, and the frequency and time delay at the minimum time delay are called the nose frequency and nose time delay respectively. The increasing time delay at higher frequencies is clear evidence that the upper frequencies cannot be considered to be much less than the electron gyrofrequency over the entire path. For whistler-mode waves propagating purely longitudinally in the magnetosphere, the propagation delay is given by

13. WHISTLERS

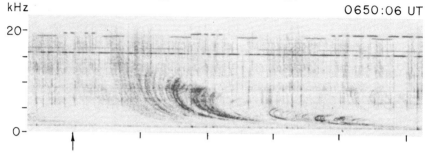
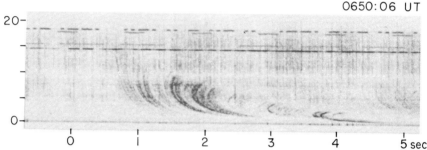

FIG. 2. Spectrograms of two nose whistler trains recorded at Eights, Antarctica. Note the repeatability of the detailed fine structure. These are one-hop whistlers with the time of origin indicated by the arrow in the upper spectrogram, and by time 0 in the lower spectrogram. [From J. J. Angerami and D. L. Carpenter, *J. Geophys. Res.* **71**, 711 (1966).]

$$t = \frac{1}{2c} \int_{\text{path}} \frac{f_p f_H}{f^{1/2}(f_H - f)^{3/2}} \, ds, \qquad (13.3.1)$$

where f_p is the electron plasma frequency, f_H the electron cyclotron frequency and c is the velocity of light in vacuum. It is readily seen that the propagation delay becomes very large both for extremely low frequencies and for frequencies close to the electron cyclotron frequency. In a uniform medium, the minimum propagation delay occurs at one-fourth of the electron cyclotron frequency. It is clear then that in the nonuniform magnetospheric medium the nose frequency provides some information about the minimum magnetic field strength along the propagation path and hence about the path location, while the propagation delay provides information about the electron density along this path. In Fig. 2, it is seen that whistlers with lower nose frequencies (corresponding to smaller minimum cyclotron frequency and hence more distant and longer propagation paths) have longer propagation delays, i.e., the nose frequency decreases with increasing time delay.

While the qualitative explanation of the nose frequency is simple, the quantitative problem of relating the nose frequency to the propagation path for observed whistlers is not so simple. It involves more detailed consideration of the properties of waves propagating in the magnetosphere and, in particular, of propagation in the magnetospheric ducts.

13.3.3. Ducting of Whistlers

Ground-based whistler observations provide overwhelming evidence for the existence of discrete ducts of enhanced ionization in the outer magnetosphere which confine the whistler energy and are responsible for the discrete components of whistlers observed on the ground. (To reduce confusion later in this paper we note that most of the energy observed in satellites in the outer ionosphere is propagating in the nonducted mode. However, this energy does not penetrate the lower ionosphere boundary and is not observed on the ground.) Part of the evidence for the existence of ducts comes from whistler echo trains, part from the discrete nature of ground-observed whistlers and part from the repeatability of the particular multicomponent structure in successive whistlers produced by different causative atmospherics. The similarity of component structure is very evident in the two whistlers shown in Fig. 2.

By identifying a particular whistler component and tracing it through successive multicomponent whistlers, some estimate of the lifetime of a magnetospheric duct is obtained. Lifetimes are found to be typically one to several hours. The repetitive nature of the component structure of ducted whistlers is extremely helpful in identifying the causative atmospheric which, for one-hop whistlers, is often difficult to detect on the spectrogram. However, the approximate location of the atmospheric may be obtained by extrapolating back from the time delay at low frequencies using the Eckersley law[4] given in Eq. (13.1.1). This limits the number of possible causative atmospherics. The actual source is then selected using the requirement that the propagation delays from the atmospheric source to the whistler components must be the same for successive multicomponent whistlers. For a given whistler, only one atmospheric will satisfy this requirement. This point is also illustrated by Fig. 2, where the time scales for the two whistlers have been aligned so that the corresponding whistler components occur at the same time. This permits the causative atmospherics to be readily identified. The ability to identify the causative atmospheric greatly improves the accuracy of measurements of propagation delay as a function of frequency.

[4] T. L. Eckersley, *Nature (London)* **135**, 104 (1935).

In order to understand how whistlers are trapped in ducts, let us consider a magnetically field-aligned enhancement of ionization for which the gradient in refractive index is normal to the magnetic field. We assume that spatial variations of the refractive index are small over a distance of many wavelengths of the wave in the medium. We represent this conceptually by a large number of parallel boundaries across which a small change in refractive index occurs. The change in direction of propagation of a wave is then determined by Snell's law, viz that the component of the propagation vector (or refractive index treated as a vector) along the boundary must be constant.† This form of Snell's law is valid for an anisotropic medium such as we are considering here.

The refractive index, n, for whistler mode waves is given by[9]

$$n = f_p/f^{1/2}(f_H \cos\theta - f)^{1/2}, \qquad (13.3.2)$$

where θ is the angle between the wave normal and the magnetic field.

Snell's law is satisfied then if $n \cos\theta$ is a constant so that

$$n \cos\theta = f_p \cos\theta/f^{1/2}(f_H \cos\theta - f)^{1/2} = \text{const.} \qquad (13.3.3)$$

A wave may be trapped in a duct if the refractive index for purely parallel propagation anywhere within the duct is smaller than the longitudinal component of refractive index for the propagating wave at the center of the duct.

For ducts of enhanced ionization, trapping of whistlers may occur if the wave normal angle at the center of the duct satisfies the condition[8]

$$\theta_0 < \theta_{0c} = \cos^{-1}(2f/f_H). \qquad (13.3.4)$$

At this angle the parallel component of refractive index, $n \cos\theta$, has a minimum, so that at larger values of θ_0, decreasing the electron density near the edge of the duct will lead to increases in wave normal angle and trapping will not occur.

[5] J. MacCullagh, *Proc. Roy. Irish Acad.* **17**, 249 (1837).
[6] W. R. Hamilton, *Proc. Roy. Irish Acad.* **17**, 144 (1837).
[7] H. Poeverlein, *Sitzber. Bayer. Akad. Wiss. Muenchen* **1**, 175 (1948).
[8] R. L. Smith, R. A. Helliwell, and I. Yabroff, *J. Geophys. Res.* **65**, 815 (1966).
[9] R. A. Helliwell, "Whistlers and Related Ionospheric Phenomena." Stanford Univ. Press, Stanford, California, 1965.

---

† The recognition of the validity of this form of Snell's law for anisotropic media usually applied with respect to "refractive index surfaces", helped accelerate understanding of whistler propagation. The importance of this form appears to have been first recognized by MacCullagh[5] and Hamilton[6] in 1837. In more recent times, Poeverlein[7] appears to have been the first to apply refractive index surfaces to ionospheric problems. Smith *et al.*[8] give a discussion relating to whistler propagation.

When (13.3.4) is satisfied, the trapping condition becomes

$$n_0 \cos \theta_0 > n_A, \quad (13.3.5)$$

$$\frac{f_{p0} \cos \theta_0}{f^{1/2}(f_H \cos \theta_0 - f)^{1/2}} > \frac{f_{pA}}{f^{1/2}(f_H - f)^{1/2}}, \quad (13.3.6)$$

where subscripts 0 and A refer to the center of the duct and the ambient medium respectively. The density enhancement required is then readily obtained as

$$\frac{N_0}{N_A} > \frac{f_H \cos \theta_0 - f}{(f_H - f) \cos^2 \theta_0}, \quad (13.3.7)$$

where N is electron density. In general, only small density enhancements are required. For example, for a wave normal angle θ_0 of 20° at the center of the duct and $f = 0.3 f_H$, the required enhancement in ionization density is only 4%.

The most important role the ducts play is the restriction of the wave normal direction of whistlers trapped in the ducts: this facilitates the transmission of energy across the lower ionosphere boundary, thus allowing part of the energy to be observed on the ground. Since the refractive index in the lower ionosphere is much greater than unity, and equal to unity below the ionosphere, total internal reflections will occur for all downcoming waves except those propagating in directions within a narrow transmission cone around the normal to the boundary. For downcoming nonducted waves, the propagation direction usually makes a large angle with respect to the magnetic field and is well outside the transmission cone for the lower ionosphere boundary.

For ducted whistlers, all frequencies in each whistler component propagate substantially along the same field-aligned path. The problem of finding the detailed whistler propagation characteristics within a duct and the resulting frequency-time dispersion of the whistler was examined by Smith.[10] For a range of frequencies, initial wave normal angles, and electron density enhancements, the ray paths within the duct were calculated. By integrating the time delay along these paths, the total propagation delay was determined. For a wide range of values of the above parameters it was found that the delays were the same (to within one or two percent) as if the wave propagated purely in the direction of the magnetic field along the center of the duct. This result greatly simplifies interpretation of the whistler dispersion in terms of electron density and magnetic field strength. It arises in part because the trapping condition initially selects a range of relatively small wave normal angles. These angles may be further

[10] R. L. Smith, *J. Geophys. Res.* **66**, 3699 (1961).

reduced by longitudinal gradients of refractive index as the wave propagates toward the equatorial plane. Also, for a given frequency, the whistler mode wave follows a snakelike path within the duct. By comparison with a hypothetical purely longitudinal whistler at the center of the duct, contributions to the propagation delay are larger when the energy is near the center of the duct and smaller when the energy is away from the center. These two effects very nearly cancel, making possible the simplifying assumption of purely longitudinal propagation.

One additional feature of propagation in ducts of enhanced ionization should be mentioned. As may be seen from Eq. (3.5), trapping cannot occur for waves at frequencies greater than half the local electron cyclotron frequency. Consider a wave propagating upward in a duct in the lower ionosphere at a frequency less than half the cyclotron frequency. As the wave propagates, the local electron cyclotron frequency decreases and the wave will escape from the duct when the (cyclotron) frequency is reduced to twice the wave frequency. If the wave frequency is less than half the minimum cyclotron frequency in the duct, it remains trapped. This leakage of higher frequency wave energy out of the duct produces an upper cutoff of ground-based whistlers at half the minimum cyclotron frequency for the path.

13.3.4. Equatorial Electron Density and Flux-Tube Content

For a given model electron density and magnetic field-line path, we may readily calculate the frequency-time behavior of ducted whistlers, by assuming for the purposes of calculation that the whistler is propagating purely longitudinally. The problem is to invert this process to obtain the parameters of the medium in terms of the experimentally measured frequency-time spectrum of a whistler, i.e., to invert the integral equation

$$t = \frac{1}{2c} \int_{\text{path}} \frac{f_p f_H}{f^{1/2}(f_H - f)^{3/2}} \, ds. \qquad (13.3.8)$$

Note that in the integration the plasma frequency and cyclotron frequency are both parameters of the path of integration, and that the integration depends on the wave frequency chosen. For simplicity, the magnetic field is approximated by a magnetic dipole. Unfortunately, it is found in practice that the frequency-time behavior, in terms of measurable quantities, is nearly independent of the plasma frequency profile along the geomagnetic line of force.

If we define the dispersion, D, as

$$D = tf^{1/2}, \qquad (13.3.9)$$

the dispersion is a function of a frequency, approaching a constant value, D_0, at low frequencies. It is found that the ratio of dispersion at the nose frequency, D_n, to dispersion at zero frequency, D_0, is essentially independent of the electron density distribution along the path of propagation. In other words if we plot f/f_n versus t/t_n, the resulting whistler shape shows no measurable change for a wide range of electron density models. From model calculations, it is found that differences do occur for frequencies near the minimum gyrofrequency. However, the properties of the ducts limit the maximum whistler frequency to half the minimum

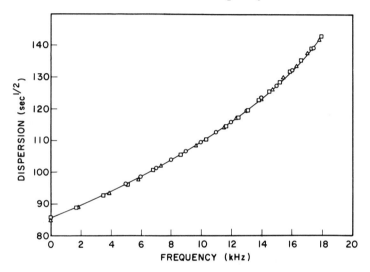

FIG. 3. Theoretical dispersion curves fitted to a whistler. The density models are considerably different, but the resulting dispersion functions, when normalized, are indistinguishable within experimental accuracy. (O) Measured from whistler, SE: February 5, 1958; (□) calculated from gyrofrequency model; (△) calculated from diffusive equilibrium model.

gyrofrequency as discussed above. This point[10] is illustrated in Fig. 3, where the dispersion is plotted as a function of frequency for a simple diffusive equilbrium model of electron density[11]

$$N \propto \exp 2.5/R \qquad (13.3.10)$$

(where R is the distance from the center of the earth in earth radii), a "gyrofrequency model"[10]

$$N \propto B, \qquad (13.3.11)$$

[11] J. W. Dungey, Sci. Rep. No. 69, Air Force Contract, No. AF19(122-33). Ionospheric Res. Lab., Pennsylvania State Univ., University Park, Pennsylvania, 1954.

and as measured from a whistler. The three curves are indistinguishable. This figure also serves to illustrate the exceptional agreement between theory and data. As a practical matter, then, there are only two independent parameters obtainable from a whistler, those most commonly used being the nose frequency, f_n, and nose time delay, t_n.

Thus, in order to obtain estimates of the electron density in the equatorial plane or the total electron tube content (above, say, 1000 km i.e., excluding the dense ionospheric region), some model for electron density along the path must be assumed. Angerami and Carpenter[12] have

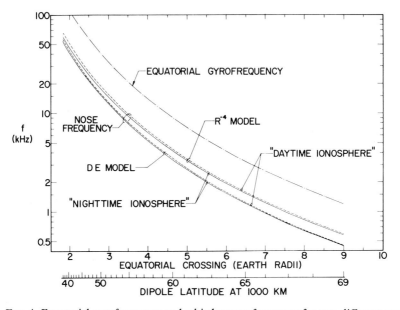

FIG. 4. Equatorial gyrofrequency and whistler nose frequency for two different magnetosphere and two different ionosphere models. (From J. J. Angerami, Ph.D. Thesis, Stanford Univ., May, 1966.)

made calculations for two models, a diffusive equilibrium (DE) model and an R^{-4} model believed to be appropriate in different regions in the magnetosphere. For the former, a temperature of 1600° K was used, and an assumed ionic composition at 1000 km altitude being 50% O^+; 10% He^+ and 40% H^+. For the latter, the electron density above 1000 km was assumed to vary as R^{-4}, where R is the geocentric distance in earth radii. For the contribution from electrons below 1000 km, a frequency-independent dispersion of $D = 6$ was added for the nighttime ionosphere with $D = 12$ for daytime. A dipole magnetic field distribution was

[12] J. J. Angerami and D. L. Carpenter, *J. Geophys. Res.* **71,** 711 (1966).

13.3. GROUND-BASED WHISTLER OBSERVATIONS

assumed. Figure 4 shows the equatorial gyrofrequency and whistler nose frequency for the two models, each with daytime and nighttime ionospheres, plotted as a function of equatorial distance and dipole latitude at 1000 km. It is seen that the ratio of nose frequency to the minimum cyclotron frequency is not sensitively dependent on the density model and is almost independent of the ionosphere.

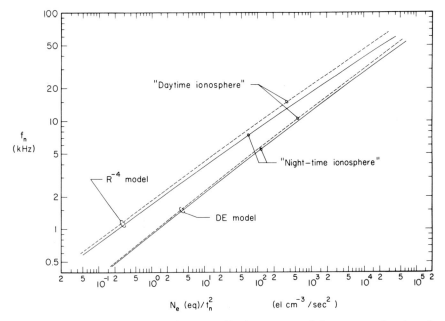

FIG. 5. Equatorial electron density, normalized to nose travel time, t_n, as a function of nose frequency, f_n. (From J. J. Angerami, Ph.D. Thesis, Stanford Univ., May, 1966.)

In Fig. 5, the nose frequency is plotted as a function of the ratio of equatorial electron density to the square of the time delay at the nose for both models. Again the ionosphere gives only a small effect except at very large nose frequencies (i.e., low altitudes). By choosing the appropriate model (see the following), the electron density can be determined directly from this curve, given the nose frequency and nose time delay. Uncertainties as to the precise electron density model make the measurements of equatorial electron density uncertain, typically by about 30%.[13–13b]

---

[13] J. J. Angerami, Tech. Rep. 3412-7. Ph.D. Thesis, Radioscience Lab., Stanford Electron. Lab., Stanford Univ., Stanford, California, 1966.
[13a] C. G. Park and D. L. Carpenter, *J. Geophys. Res.*, **75**, (1970).
[13b] R. A. Helliwell, *Rev. Geophys.*, **7**, 281 (1969).

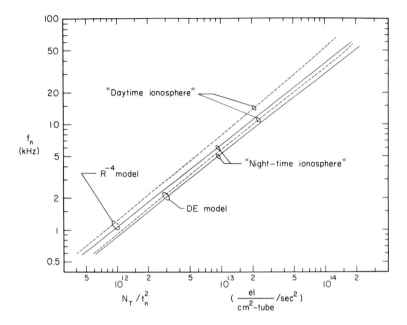

FIG. 6. Integrated electrons in a tube of magnetic flux as a function of nose frequency. (From J. J. Angerami, Ph.D. Thesis, Stanford Univ., May, 1966.)

In Fig. 6, the ratio of tube content divided by nose time delay squared is plotted as a function of nose frequency for the two models. "Tube content" as defined here refers to the total number of electrons contained in a magnetic flux tube of cross section 1 cm² at 1000 km, between 1000 km and the equatorial plane. From this figure the tube content can be found from the nose frequency and nose time delay. The role played by the ionosphere below 1000 km is more important in these measurements. However, the estimate of tube content obtained is less sensitive to variations of the electron density distribution above 1000 km than is the equatorial density estimate, the uncertainty in the measurement of tube content being about 20%.[13-13b]

Multicomponent whistlers covering the same range of altitudes but at different longitudes will measure different electron densities if there are longitudinal gradients in density. Since a single whistler station receives whistlers from a range of longitudes (about ±15°), whistlers received during one period of time at one station may show, for a given nose frequency, a range of nose time delays (i.e., a range of electron densities). Electron density measurements which were made from whistlers recorded at Eights Station in Antarctica and which clearly demonstrate the

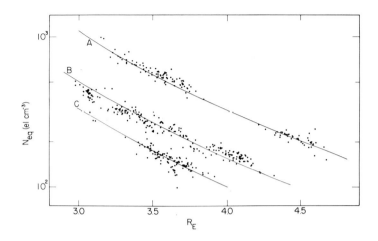

FIG. 7. Equatorial electron densities deduced from whistlers observed at Eights, Antarctica, June 9, 1965, 0150–1040 UT; R_E is the equatorial distance from the center of the earth, expressed in earth radii. [From C. G. Park and D. L. Carpenter, *J. Geophys. Res.*, **75**, 7 (1970).]

existence of strong longitudinal gradients in density are shown in Fig. 7. In this instance, the observed electron densities were associated with three distinct electron density profiles, labeled A, B, and C in Fig. 7, indicating that whistlers from three distinct longitude regions were being received at Eights Station during this period.

13.3.5. Knee Whistlers

For the whistlers shown in Fig. 2, the nose frequency decreases continuously with increasing time delay. This corresponds to a continuous gradual decrease in equatorial plane electron density with increasing distance, as illustrated by the three profiles in Fig. 7. However, **particularly at high latitudes**, the whistler spectrogram is as illustrated in Fig. 8a, in which as the nose frequency decreases, the nose time delay initially increases, then decreases abruptly. The corresponding equatorial electron density profile (Fig. 8d) exhibits an abrupt decrease in the electron density profile which Carpenter[14] referred to as a "knee." The density decreases typically from a few hundred to a few tens per cubic centimeter in a fraction of an earth radius. Whistlers showing this characteristic are called "knee" whistlers. The location of the abrupt density decrease is

[14] D. L. Carpenter, *J. Geophys. Res.* **68**, 1675 (1963).

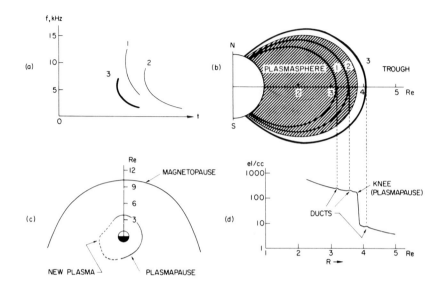

FIG. 8. (a) Idealized nose whistler train with knee trace; (b) meridional plane showing whistler ducts; (c) boundary of plasmapause deduced from knee whistlers; (d) equatorial profile of electron density deduced from train shown in (a). [From R. A. Helliwell, *Rev. Geophys.* **7**, 281 (1969).]

also called the "plasmapause." Figure 8b is a section of a meridional plane showing the location of the three whistler ducts, with the shading indicating the region of high plasma density known as the "plasmasphere."

Figure 9 shows examples of spectrograms of two knee whistlers recorded at Eights Station. The upper two spectrograms show the same whistler, displayed with two different frequency scales, while the lower whistler was recorded one minute later. In each figure, the first nose whistler trace is the whistler outside the plasmapause, the later traces being from ducts inside the plasmapause. These observations were made at times of moderate geomagnetic activity, the K_p index varying between 2 and 4.

The whistler technique is well suited to measure the location of the plasmapause. It has been shown that the average geocentric distance to the plasmapause in the equatorial plane is about $4R_E$. The plasmapause is found at smaller equatorial distances during geomagnetically disturbed periods (as low as $2R_E$ during very disturbed periods) and at larger distances during quiet periods.[15] Whistler evidence suggests that the

[15] D. L. Carpenter, *J. Geophys. Res.* **71**, 693 (1966).

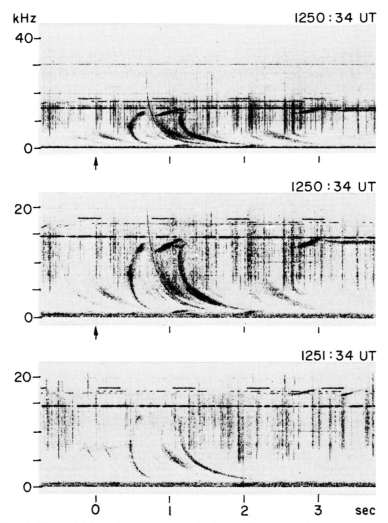

FIG. 9. Knee whistlers observed during the day, ∼0750 LT, July 7, 1963, at Eights, Antarctica. The arrow and zero second mark indicate the time of the causative atmospheric. [From D. L. Carpenter, *J. Geophys. Res.* **71**, 693 (1966).]

plasmapause is a permanent feature of the magnetosphere. A diurnal variation of the equatorial location of the plasmapause as measured from whistlers by Carpenter[15] is shown in Fig. 8c, which also indicates the location of the outer boundary of the magnetosphere or magnetopause. In steady state models of the magnetosphere, the plasmapause is believed to represent the boundary between magnetic flux tubes whose motion is dominated by corotation of the earth and those whose motion is primarily convective, driven by the flow of solar plasma past the earth (the "solar wind").[16,17]

Occasionally whistlers provide density information at large distances beyond the knee, particularly in the early afternoon hours. Figure 10

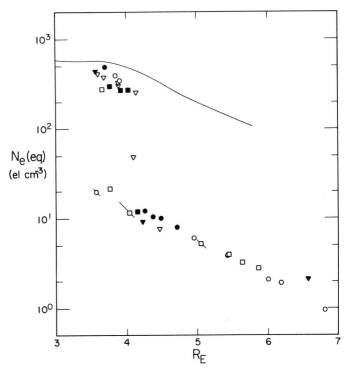

FIG. 10. Three examples of typical afternoon equatorial profiles in which the plasmapause is approximately at 4 earth radii. Note the repeatability of density levels on both sides of the plasmapause. The continuous curve is a reference for extremely low geomagnetic activity. (■□) July 31, 1963: 2150, 2210 UT (1650, 1710 LT); (▼▽) August 2, 1963: 1750 UT (1250 LT); (●○) August 4, 1963: 1950 UT (1450 LT). [From J. J. Angerami and D. L. Carpenter, *J. Geophys. Res.* **71**, 711 (1966).]

shows electron density as a function of distance in the equatorial plane deduced from whistlers recorded on 3 different days at Eights Station. The smooth curve is a reference profile obtained during a magnetically very quiet period. Solid symbols represent measurements of the highest precision, open symbols indicating those of less accuracy, while the

[16] A. Nishida, *J. Geophys. Res.* **71**, 5669 (1966).
[17] N. M. Brice, *J. Geophys. Res.* **72**, 5193 (1967).

least accurate points have bars attached indicating the range of uncertainty.

Figure 11 shows measurements of total tube content for two different local time regions, the closed circles representing near midnight and early morning conditions and the open circles early afternoon. It is interesting to note that outside the plasmapause the daytime tube contents are larger than nighttime ones, by a factor of 2 or 3. This result supports the suggestion[17] that empty flux tubes are swept from the open field line

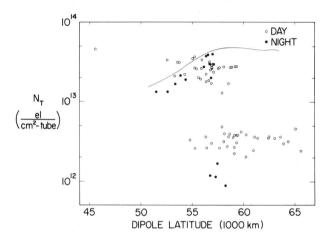

FIG. 11. Day (○) and night (●) comparison of electron tube content in the magnetosphere on both sides of the plasmapause. The continuous curve is a reference for extremely low geomagnetic activity. [From J. J. Angerami and D. L. Carpenter, *J. Geophys. Res.* **71**, 711 (1966).]

"tail" region on the night side through the closed field-line body of the magnetosphere to the day side, with the tube slowly being filled with plasma from the ionosphere in the closed field-line region. Fluxes of plasma flowing up out of the ionosphere (the "polar wind") calculated by Banks and Holtzer[18] are consistent with the day–night difference in tube content shown in Fig. 11.

13.3.6. Electron Density Distribution along Field Lines

It was noted above that the whistler gives a weighted integral measure of the electron density, with the principal uncertainty in equatorial electron density or tube content arising from uncertainties in the actual electron density distribution. These uncertainties can be reduced both

[18] P. M. Banks and T. E. Holtzer, *J. Geophys. Res.* **73**, 6846 (1968).

by physical reasoning (e.g., as to the applicability or otherwise of a field-aligned diffusive equilibrium distribution), and by additional data (e.g., the electron density distribution as a function of latitude at 1000 km measured by the Alouette I and II satellites).

Information about a change in distribution from inside to outside the plasmapause has been obtained from whistlers by Angerami.[13] He found that the nose frequency of the first whistler trace outside the plasmapause was often slightly higher than for the last whistler inside the plasmapause. The effect persisted at times of day when longitudinal gradients in density and in the location of the plasmapause tended to be small.

Since the minimum electron gyrofrequency must be smaller on the whistler trace outside the knee, the data clearly showed that the ratio of nose frequency to minimum electron gyrofrequency was larger outside the plasmapause than inside. The difference in this ratio was found to be consistent with the difference between a diffusive equilibrium distribution inside the plasmapause and an R^{-4} distribution outside the plasmapause.[13]

If the upper cutoff frequency of the whistler is known to be accurately 0.5 of the minimum gyrofrequency, then, from the measured ratio of nose frequency to upper cutoff frequency, the ratio of nose frequency to minimum gyrofrequency can be obtained. This measurement presumably can supply additional information about the distribution of electron density along the path.

13.3.7. Electric Field Measurements

As the plasma in the ionosphere is an excellent conductor, the flow of plasma in the magnetosphere with velocity **v** may be related to transverse electric fields **E** by the relationship

$$\mathbf{E} + \mathbf{v} \times \mathbf{B} = 0. \tag{13.3.12}$$

For steady flow normal to the magnetic field direction, the flow lines are electrostatic equipotentials. The plasmapause is believed to be a flow line for the magnetospheric plasma and the diurnal variation of the plasmapause location has been used to estimate the quasistatic average electric fields associated with convective flow in the magnetosphere.[17]

Whistlers have also been used to measure transient large-scale electric fields, by following the movement of ducts in which the whistlers propagate.[19] Figure 12 shows an example using whistlers recorded at Eights Station in Antarctica. The upper traces in this figure show the magneto-

---

[19] D. L. Carpenter and K. Stone, *Planet. Space Sci.* **15,** 395 (1967).

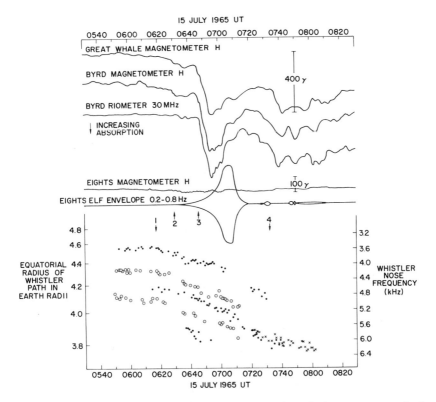

FIG. 12. Large-scale electric fields in the magnetosphere during a magnetospheric substorm as indicated by whistlers, magnetometers, and riometers. [From D. L. Carpenter and K. Stone, *Planet. Space Sci.* **15**, 395 (1967).]

meter (horizontal component) data from Byrd Station in Antarctica and its conjugate station, Great Whale River in Canada, and the riometer trace (measuring ionospheric absorption of cosmic radio noises) at Byrd Station. In the lower part of the figure are shown the nose frequencies of individual whistler components and the corresponding geocentric equatorial distances measured in earth radii and referred to as L value. The upper traces show a sudden decrease in horizontal magnetic field strength at Byrd and Great Whale River and a sudden increase in the absorption of cosmic noise (the riometer trace showing absorption as increasing downward) indicating the existence of a "magnetospheric substorm."[20,21] Some time before the

[20] D. Jelly and N. M. Brice, Astron. Center No. 90, Cornell-Sydney Univ., Ithaca, New York, 1967.
[21] N. M. Brice, in "Physics of the Magnetosphere" (R. L. Carovillano, ed.), p. 563. Reidel, Dordrecht, Holland, 1968.

onset of substorm activity at Byrd and Great Whale River, a general gradual increase in the nose frequency of whistler components begins. This is caused by a gradual inward movement of the ducts of enhanced ionization in which the whistlers propagate. From the change in whistler nose frequency with time, the radial component of velocity may be determined and from this the azimuthal electric fields. These inward movements are associated with inward movement of the plasmapause and are consistent with the hypothesis that these electric fields are enhancements of the average large-scale convective electric fields in the magnetosphere. These enhanced electric field values are typically a few kilovolts per earth radius.

Whistler studies have produced other interesting evidence related to electric fields and plasma flow in the magnetosphere. The afternoon "bulge" which is shown on the dusk side of the earth in Fig. 8c is believed to result from a "standoff" between the convective flow toward the day side of the earth and corotational flow toward the night side. Sudden decreases in convective electric field would allow the corotational flow to dominate in the bulge region, so that the bulge would begin to move into the night hours. Sudden increases in the convective field cause rapid increases in the inward motion of the plasmapause in the early morning hours and may cause outward motion in the late morning hours. These and other fascinating complex phenomena related to magnetospheric motions and electric fields have been observed using whistlers. An indication of the wide variety of effects deduced from whistler observations at different local times is given by Fig. 13.

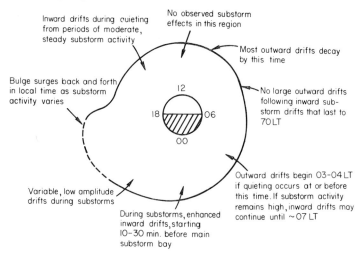

FIG. 13. The relation of the plasmapause to large-scale magnetospheric drifts. (From D. L. Carpenter, Private communication, 1970.)

13.4. Satellite Observations

13.4.1. Ducted Whistlers

Several earth-orbiting satellites have carried broad-band whistler receivers, and have detected ducted whistlers. As these whistlers are only received when the satellite is in the duct, or below it, if the duct terminates at a large altitude, the number of such whistlers received is relatively small. Nonducted whistlers (see the following) are much more plentiful. Ducted whistlers observed in satellites provide useful information about the *distribution* of ionization along the duct, information which is not obtainable from ground stations.[22] This point is illustrated in Fig. 14.[22] If a satellite

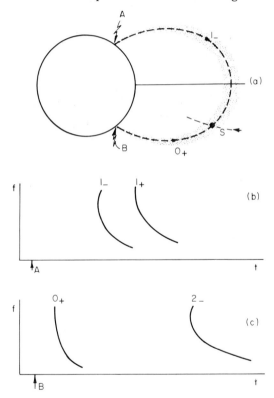

FIG. 14. Sketch showing nomenclature of ducted whistlers observed in a high altitude satellite (S). The duct is indicated by the dotted region in (a). A source in the same hemisphere as the satellite (B) produces 0_+ and 2_- whistlers, shown in (c), while a source in the opposite hemisphere produces 1_- and 1_+ whistlers, as shown in (b). [From R. L. Smith and J. J. Angerami, *J. Geophys. Res.* **73**, 1 (1968).]

[22] R. L. Smith and J. J. Angerami, *J. Geophys. Res.* **73**, 1 (1968).

is in, say, the southern hemisphere, it may receive whistlers from a northern hemisphere lightning discharge (labeled (A) in Fig. 14) after the whistler has crossed the equatorial plane (the 1_- whistler)† and again after the whistler is reflected from the southern hemisphere ionosphere (the 1_+ whistler).† A southern hemisphere source (B) will give a 0_+ whistler and a 2_- whistler. These give information about the propagation delay (and hence electron density) along a part of total path, from which information is obtained about the density in the outermost portion of the path relative to that nearer the earth. The 0_+ whistler also may be expected to show an upper-frequency cutoff near half the electron cyclotron frequency at the satellite. Waves at higher frequencies cannot remain trapped and leak out of the duct. Because of the curvature of the magnetic field toward the earth, the wave normal becomes pointed in a direction away from the earth with respect to the magnetic field, after the waves leak out of the duct. For frequencies above half the cyclotron frequency because of the anisotropy of the medium, an outward wave normal causes the wave energy to move inward, i.e., toward the earth (again with respect to the magnetic field). Thus a satellite approaching the earth may record a ducted whistler at frequencies below half the cyclotron frequency in a duct. At a later time when the satellite is nearer the earth, it may receive wave energy at higher frequencies which have leaked out of the duct. This effect is illustrated in Fig. 15, which shows two ducts.

A ground-based recorder receives whistlers propagating in the two ducts (upper spectrogram) with each whistler showing an upper cutoff at half the minimum cyclotron frequency for the path. The satellite in the *inner* duct (1) receives a 0_+ whistler which has an upper cutoff above half the minimum gyrofrequency (and at approximately half the *local* gyrofrequency) and the leakage from the outer duct (2). This is illustrated in the lower right-hand part of Fig. 15.

13.4.2. Ion Effects

13.4.2.1. The Subprotonospheric Whistler. In all of the preceding discussion it has been implicitly assumed that the wave frequencies are much higher than the ion cyclotron frequency so that ion currents could be ignored.

The importance of ion currents was indicated by observations of subprotonospheric (SP) whistlers on the Alouette satellite at 1000 km

---

† The nomenclature here results from extension of nomenclature based on ground observations to satellite observations. Thus a 1_- whistler is nearly a one-hop whistler which has made one crossing of the magnetic equator but has not yet suffered a reflection.

13.4. SATELLITE OBSERVATIONS

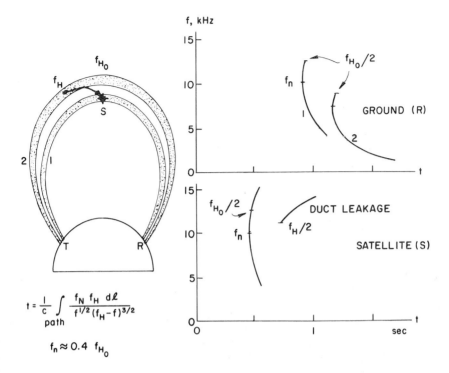

FIG. 15. Multipath ducted nose whistlers observed on the ground (R) and satellite (S). The source is at T. The ground observes an upper-frequency cutoff of each component at half the minimum gyrofrequency for each path. The satellite may observe only one ducted component at a time, plus higher frequency components which are about to depart from the ducted path, and other leakage components which have already propagated away from other ducts. [From R. A. Helliwell, *Rev. Geophys.* **7**, 281 (1969).]

altitude (see Barrington and Belrose[23]) and a rocket in the lower ionosphere (Carpenter *et al.*[24]). An example of an Alouette SP whistler is shown in Fig. 16. This spectrogram shows a short fractional hop (or 0^+) whistler with dispersion D followed by other traces with dispersion $3D$, $5D$, etc. The rocket receiver in the lower ionosphere showed SP whistler traces with dispersion 0, $2D$, $4D$, etc.

It was apparent that these whistlers were refracted from upgoing to downcoming in the topside ionosphere (probably in the height range 1000–3000 km) and reflected back up from the lower ionosphere (or perhaps the earth). Smith[25] showed that the refraction from upgoing to

[23] R. E. Barrington and J. S. Belrose, *Nature (London)* **198**, 651 (1963).
[24] D. L. Carpenter, N. Dunckel, and J. F. Walkup, *J. Geophys. Res.* **69**, 5009 (1964).
[25] R. L. Smith, *J. Geophys. Res.* **69**, 5019 (1964).

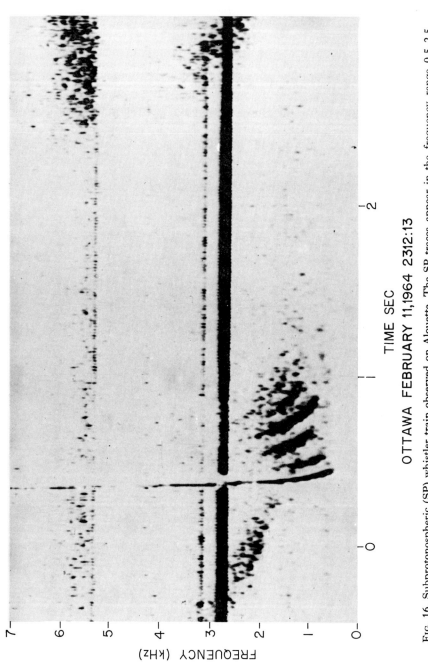

FIG. 16. Subprotonospheric (SP) whistler train observed on Alouette. The SP traces appear in the frequency range 0.5–2.5 kHz. Note the very small amount of dispersion compared to normal whistlers.

downcoming could not be explained unless ion currents were taken into consideration. In this case, Eq. (13.3.2) (which is derived ignoring ions and indicates that propagation perpendicular to the field is not possible) becomes modified at frequencies below that of the lower hybrid resonance (LHR) so that propagation is possible for all directions. The SP whistler provides some (not very specific) information about ions since all the frequencies in SP whistlers (excluding the first short fractional hop) must be below the LHR frequency. Subprotonospheric whistlers were observed on the ground in 1958 by Carpenter[25a] but not investigated in detail. The dispersion of these whistlers gives a measure of the total electron content in the ionosphere between the earth and the height of reflection.

13.4.2.2. The Magnetospherically-Reflected Whistler. Since the refractive index in the ionosphere is very large (~ 20) at whistler frequencies, the

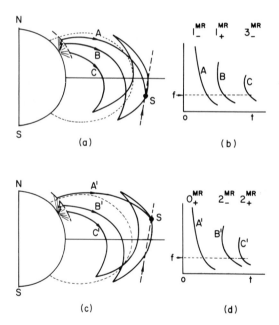

FIG. 17. Magnetospherically reflected (MR) whistlers. The source illuminates a large region of the ionosphere. The ray path which passes through the satellite is a function of frequency and hop number. When the source is in the opposite hemisphere from the satellite, the spectrum has the appearance illustrated in (b). A typical set of ray paths for frequency f is shown in (a). For the source in the same hemisphere, the spectrum is illustrated in (d), and the ray paths in (c). [From R. L. Smith and J. J. Angerami, *J. Geophys. Res.* **73**, 1 (1968).]

[25a] D. L. Carpenter, Private communication, 1970.

wave normal angle of waves entering from below is essentially parallel to the gradient in electron density, i.e., nearly vertical. As a result most of the energy is *not* refracted from upgoing to downcoming in the upper ionosphere (to produce SP whistlers) but propagates out across the magnetic equator and down in the opposite hemisphere. For nonducted waves, the path taken by the wave may be determined in a known electron and ion distribution by ray tracing.[22] An example showing calculated 1_-, 1_+, and 3_- ray paths for nonducted whistlers is shown in Fig. 17a. Figure 17b shows the calculated spectrograms of the 1_-, 1_+, and 3_- nonducted whistlers. A fine example of an actual nonducted whistler is shown in Fig. 18. From these whistlers it is possible to obtain the electron and ion

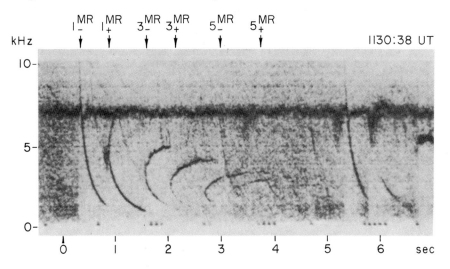

FIG. 18. Spectrogram of MR whistler observed on the OGO-1 satellite, November 8, 1965. [From R. L. Smith and J. J. Angerami, *J. Geophys. Res.* **73**, 1 (1968).]

density distributions in the ionosphere and magnetosphere by ray tracing and adjusting the model until a fit is obtained to the observed whistlers. Because of the high refractive index in the ionosphere, the initial wave normal angle at the base of the ionosphere is assumed to be vertical. For a given frequency and given hop number, the point at which the wave must enter the ionosphere in order to reach the satellite must be determined by trial and error. Because the required computations are rather lengthy, these nonducted magnetospherically reflected (MR) whistlers are not used on a regular basis for determining the magnetospheric density distribution. If the down-coming and up-going waves are recorded near the reflection point, they yield information about the ion and electron distribution near the reflection point with more accuracy and less difficulty than for whistlers away from the reflection point.

13.4.2.3. Lower Hybrid Resonance Noise.

Barrington and Belrose[23] reported finding in the Alouette records an unusual noise band with a sharp lower cutoff frequency, which consistently increased with decreasing latitude of the satellite. The frequency of the noise band was well above the proton cyclotron frequency and well below the electron cyclotron frequency and it soon became apparent that the lower cutoff frequency was the lower hybrid resonance (LHR) frequency.[26,27] The change in frequency with latitude was explained by Brice[28] as being due partly to a change in electron density and partly to an increase of the fractional abundance of oxygen ions (or decrease in hydrogen ions) with latitude at 1000 km, the height of the satellite. An example of LHR noise is shown in Fig. 19. In this example, the LHR noise appears to be "stimulated" by both upgoing 0_+ and downcoming 2_- whistlers. The noise which follows the 2_- whistler trace is believed to be energy from the lightning source propagating in the "nonducted mode." Ray path calculations such as those used for the magnetospherically reflected waves show that for nonducted propagation, large wave normal angles are generally expected. One can think of this in terms of the magnetic field direction curving inward away from the wave normal direction, giving a wave normal direction which is outward with respect to the magnetic field direction. For frequencies above the lower hybrid resonance, as the wave normal angle becomes large, it approaches the resonance cone angle and the component of refractive index normal to the field becomes very large. These waves are reflected when the wave frequency passes below the LHR frequency and the perpendicular component of refractive index is no longer infinite and decreases rapidly with decreasing height. For example, if the nonducted whistler wave is propagating with a wave normal angle very close to the resonance cone angle, then the refractive index will be very large, say 600. For frequencies much less than the electron gyrofrequency, but above the LHR, the resonance cone angle is near 90°, so the component of refractive index normal to the field would be essentially the same as the refractive index (600 in this case).

As the LHR frequency increases along the path, it becomes larger than the wave frequency, and propagation becomes possible across the field. At the height were the refractive index for propagation normal to the

---

[26] N. M. Brice and R. L. Smith, *Nature (London)* **203**, 926 (1964).

[27] N. M. Brice and R. L. Smith, *J. Geophys. Res.* **70**, 71 (1965).

[28] N. M. Brice, Rep. No. SEL 64-088. Radioscience Lab., Stanford Electron. Lab., Stanford Univ., Stanford, California, 1964.

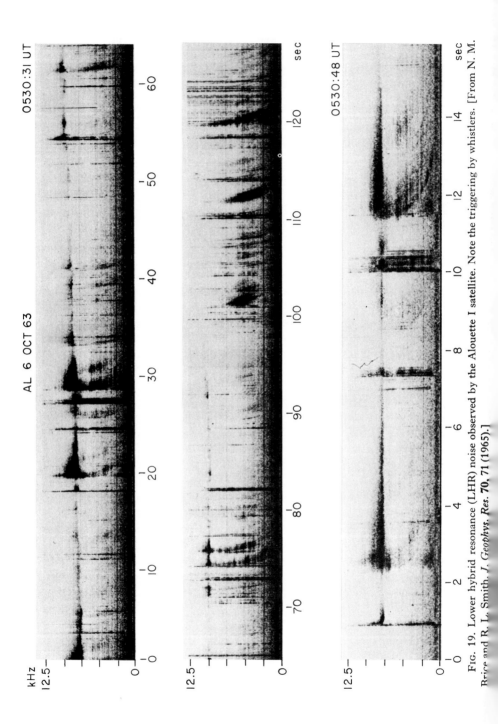

FIG. 19. Lower hybrid resonance (LHR) noise observed by the Alouette I satellite. Note the triggering by whistlers. [From N. M. Brice and R. L. Smith, *J. Geophys. Res.* **70**, 71 (1965).]

field becomes equal to 600 (in this case), the wave is reflected. In duct propagation previously considered, we considered gradients in the parameters of the medium *normal* to the field for which Snell's law requires the *parallel* component of refractive index to be constant. Here we consider changes in the medium parameters *along* the magnetic field direction so from Snell's law the *normal* component of refractive index must remain constant. Because the refractive index for propagation normal to the field decreases very rapidly as the frequency is decreased below the LHR, a refractive index as large as 600 would occur only at a frequency very close (within about 1%) to the LHR frequency. Waves at large wave normal angles at higher frequencies will pass through the LHR frequency and be reflected at lower heights but may also be observed by the satellite.

Thus the upper-cutoff frequency of the noise band is generally not very well defined. The amplitude of the noise appears to be largest at the lower cutoff frequency and to decrease with increasing frequency. The best estimate of the LHR frequency is the well-defined lower cutoff frequency observed in the noise band and the accuracy of this estimate is extremely good.

As noted above, it was soon recognized that the magnitude of the variation of lower hybrid resonance frequency with latitude could not be explained by variation in electron density alone, but implied a change in fractional ion abundance as a function of latitude. In the presence of multiple ion species (as in the ionosphere) the LHR frequency, f_{LHR}, is given approximately by[26]

$$\sum_{\text{ions}} \frac{A_i}{M_i} \frac{1}{f_{\text{LHR}}^2} = \frac{1}{f_p^2} + \frac{1}{f_H^2}, \qquad (13.4.1)$$

where each A_i is the fractional (numerical) ion abundance and M_i is the ratio of the ion to electron mass. It is seen that for a multiple ion plasma, the ion mass is replaced by an effective mass $M_{i\,\text{eff}}$ which is the (weighted) harmonic mean ion mass,

$$1/M_{i\,\text{eff}} = \sum_{\text{ions}} A_i/M_i. \qquad (13.4.2)$$

It should be noted that the effective mass is dominated by the light ion abundance in the same way as the average mass is dominated by the heavy ion abundance. The discovery and explanation of the LHR noise band led to consideration of multiple-ion effects in wave propagation[29] and to the explanation and use of a number of multiple-ion related

---

[29] R. L. Smith and N. M. Brice, *J. Geophys. Res.* **69**, 5029 (1964).

phenomena described in the following. Lower hybrid resonance noise has been used to measure the effective mass in the ionosphere.[30,31]

13.4.3. Multiple Ion Effects

13.4.3.1. The Proton Whistler and the Relative Abundance of Protons. A phenomenon observed in broad band satellite recordings which has supplied a very large amount of information about the propagation medium is the proton whistler, a spectrogram of a proton whistler recorded by Injun III satellite being shown in Fig. 20. The proton whistler trace breaks away from the short fractional hop (electron, O_+) whistler at a frequency designated f_{12} in Fig. 20 and called the crossover frequency. The proton whistler frequency then increases, asymptotically approaching the proton gyrofrequency at the satellite. It is interesting that the first examples detected (aurally) were ignored since the proton gyrofrequency for low altitude satellites is close to a commonly used converter frequency (400 Hz) and the sound of a proton whistler resembled that of transient interference due to a motor being switched on and attaining synchronous speed. When the same "motor interference" was heard in two different satellites, the noise was soon realized to be of natural origin. The proton whistler appeared to be a left-hand polarized wave which was the proton equivalent of the right-hand polarized (electron) whistler trace above the nose frequency except that the proton whistler showed no upper cutoff corresponding to the duct cutoff in electron whistlers at half the minimum gyrofrequency.

In the lower ionosphere F and E regions, where the ion constituents are O^+, and O_2^+ and NO^+, respectively, left-hand polarized waves can only propagate at extremely low frequencies, i.e., at less than the ion cyclotron frequencies (about 40 Hz and 20 Hz, respectively). Thus the left-hand polarized waves observed at frequencies of about 400 Hz in the upper ionosphere (500–2500 km) must have propagated through the lower ionospheric E and F regions as right-hand polarized waves.

Examination of the theory of wave propagation with multiple-ion species showed that as the wave frequency moved from below to above a crossover frequency the polarization changed from right-circular to left-circular or vice versa, with all waves having linear polarization at the crossover frequency.[29] Associated with the change of polarization is a

---

[30] R. E. Barrington, J. S. Belrose, and G. L. Nelms, *J. Geophys. Res.* **60**, 1647 (1965).

[31] T. Laaspere, M. G. Morgan, and W. C. Johnson, *J. Geophys. Res. Space Phys.* **74**, 141 (1969).

[31a] Neil Brice, "Electromagnetic Wave Theory, Part I." Pergamon Press, Oxford and New York, 1967.

13.4. SATELLITE OBSERVATIONS

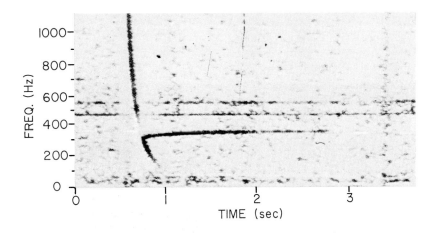

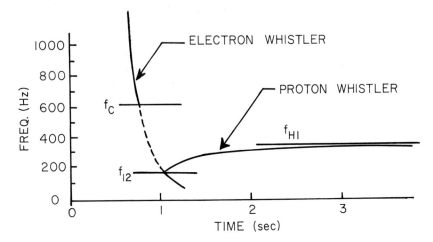

FIG. 20. Proton whistler spectrogram and associated nomenclature. [From D. A. Gurnett et al., J. Geophys. Res., **70**, 1665 (1965).]

change of group velocity, the group velocity being much smaller for left-hand polarized waves. This point is illustrated by Fig. 21, which shows the wave frequency f (normalized by the proton cyclotron frequency f_{H1}) plotted as a function of wave number, k, for wave propagating parallel to the magnetic field in a plasma containing electrons and hydrogen, helium, oxygen ions. In this diagram, k goes to zero for the left-hand mode at ion–ion cutoff frequencies and becomes infinite at the helium and

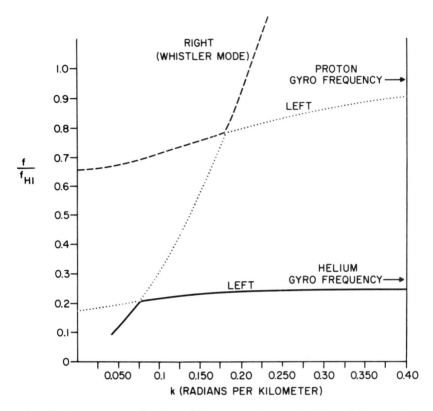

FIG. 21. Frequency as a function of the propagation constant ($\omega - k$ diagram) for a plasma containing electrons, protons, helium and oxygen ions. The lines indicate how the modes are joined for propagation at directions other than strictly along the field lines, in the absence of collisions. The proton whistler results from propagation indicated by the dotted lines. (From Neil Brice, "Electromagnetic Wave Theory, Part I." Pergamon Press, Oxford and New York, 1967.)

proton gyrofrequencies. Frequencies at which the left- and right-hand polarized modes have the same k (and hence the same phase velocity) are crossover frequencies.

For waves propagating in a region where the cyclotron frequency is decreasing (upgoing waves) the ratio f/f_{H1} is increasing and may move from below a crossover frequency to above. When this occurs, continuity is maintained from the fast mode (i.e., higher phase velocity) below the crossover frequency to the fast mode above the crossover frequency (i.e., from left-hand polarized to right-hand polarized) or from the slow mode to the slow mode (i.e., right-hand polarized to left-hand). This continuity is indicated by the symbols used in Fig. 21.

Note that at the crossover frequency, the slope of the $\omega - k$ curve (i.e., the group velocity) is much smaller for the left-hand wave than for the right-hand wave so that waves have an abrupt change in group velocity as the crossover frequency moves from below the wave frequency to above, or vice versa. For proton whistlers, we are concerned with the slow mode between the helium and hydrogen gyrofrequencies (the dotted curve in Fig. 21). As the crossover frequency passes from above the wave frequency to below, the polarization changes from right-handed to left-handed with a resultant large decrease in group velocity. The discontinuity in group velocity at the crossover frequency produces a discontinuity in the slope of the frequency-time spectrogram of the proton whistler. Thus the frequency at which the proton whistler trace departs from the electron whistler is the crossover frequency and separates the lower frequency right-hand polarized waves from the higher frequency left-hand polarized waves of the proton whistler.[32]

Crossover frequencies occur between adjacent ion gyrofrequencies and are given[33] by

$$\sum_{\text{ions}} A_i/(f_H^2 - M_i^2 f^2) = 1/f_H^2 \qquad (13.4.3)$$

The crossover frequencies are independent of electron density, depending only on the electron cyclotron frequency, f_H, the fractional ion abundances, A_i, and the ion to electron mass ratios, M_i.

For three ions, hydrogen, helium and (atomic) oxygen, of masses 1, 4, and 16 amu, this equation may be inverted and put in the form[34]

$$A_i = \frac{4^2 16^2}{(4^2 - 1)(16^2 - 1)} (1 - \Lambda_{12}^2)(1 - \Lambda_{23}^2) \qquad (13.4.4)$$

where Λ_{12} and Λ_{23} are the ratios of proton-helium and helium-oxygen crossover frequencies to the proton gyrofrequency, and A_i is the fractional abundance of protons.

Since the lower crossover frequency is between the helium and oxygen cyclotron frequencies, possible values of Λ_{23} range from $\frac{1}{4}$ to $\frac{1}{16}$, so that $1 - \Lambda_{23}^2$ has a maximum range from approximately 0.94 to 1.00 and can be accurately approximated by 0.97. Then from a knowledge of the hydrogen-helium crossover frequency (Λ_{12}) alone, the abundance of hydrogen may be obtained with an uncertainty of only $\pm 3\%$. This is demonstrated in Fig. 22 where the fractional abundance of hydrogen ions is plotted

---

[32] D. A. Gurnett, S. D. Shawhan, N. M. Brice, and R. L. Smith, *J. Geophys. Res.* **70**, 1665 (1965).
[33] N. M. Brice, *Radio Sci.* **69**, 257 (1965).
[34] S. D. Shawhan and D. A. Gurnett, *J. Geophys. Res.* **71**, 47 (1966).

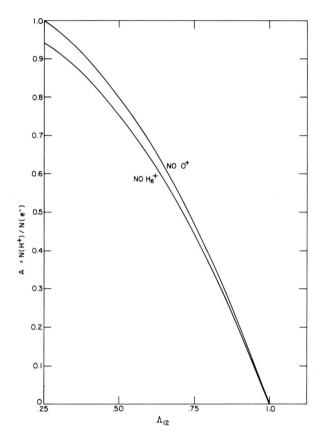

FIG. 22. Fractional abundance of protons as a function of the crossover frequency, f_{12}, for the case of no oxygen or no helium ions.

against the normalized crossover frequency Λ_{12} for H^+–He^+ and for H^+–O^+ mixtures.

Measurements of the crossover frequency obtained from the satellite Injun III have been used[34] to obtain the fractional abundance of hydrogen as a function of altitude for summer day and winter night conditions, the results being shown in Fig. 23. From these data, model distributions for the proton abundance versus height at different latitudes were obtained as shown in Fig. 24.

13.4.4. Coupling Phenomena

One would expect on the basis of the previous discussion that all short fractional hop whistlers would exhibit proton whistler characteristics. In

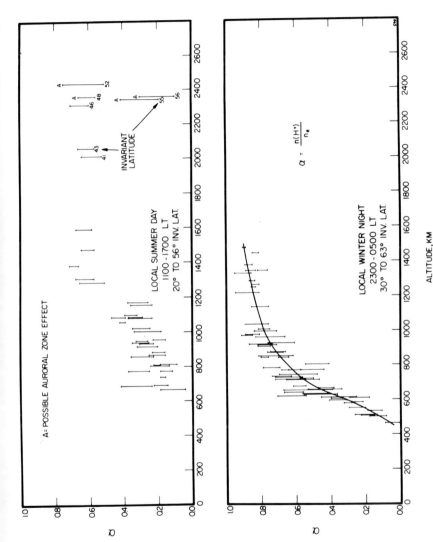

Fig. 23. Abundance of protons versus altitude for summer day and winter night. [From S. D. Shawhan and D. A. Gurnett, *J. Geophys. Res.* **71**, 47 (1966).]

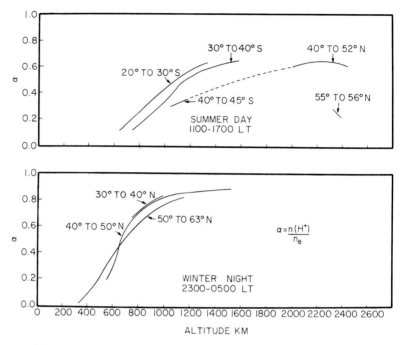

FIG. 24. Abundance of protons versus altitude for summer day and winter night. Data is separated by invariant latitude in 10° increments. [From S. D. Shawhan and D. A. Gurnett, *J. Geophys. Res.* **71**, 47 (1966).]

addition, the electron whistler portion would be expected to show a gap in its energy spectrum from the crossover frequency up to a frequency corresponding to the cyclotron frequency at an altitude below the satellite where hydrogen becomes an important constituent. However, many short fractional hop whistlers do not have proton whistler components. Furthermore, many whistlers containing proton whistler components show only a negligible decrease in energy in the frequency range above the crossover frequency (see Fig. 27 for example). This is evidence of coupling of the slow mode to the fast mode in the vicinity of the crossover frequency. In the absence of collisions, coupling is allowed only for propagation strictly in the direction of the static magnetic field. However, when collisions are taken into account, coupling between the modes is allowed over a range of directions of propagation. Furthermore, partial coupling may exist depending not only on the propagation direction but on the gradients of refractive index.[32,35,36]

[35] D. A. Gurnett and N. M. Brice, *J. Geophys. Res.* **71**, 3639 (1966).
[36] D. Jones, *J. Atmos. Terr. Phys.* **31**, 971 (1969).

13.4.5. The Proton Whistler and Magnetic Field Strength

For the proton whistler waves propagating longitudinally near the proton cyclotron frequency in the ionosphere, the refractive index, n, is given (to a very good approximation) by

$$n = f_{p1}/f_{H1}^{1/2}(f_{H1} - f)^{1/2} \tag{13.4.5}$$

and the group velocity, W_g, by

$$W_g = 2c(f_{H1} - f)^{3/2}/f_{p1}f_{H1}^{1/2}, \tag{13.4.6}$$

where f_{p1} and f_{H1} are the proton plasma and cyclotron frequencies respectively. For waves propagating upwards into a region of decreasing magnetic field strength, most of the propagation delay occurs very near the satellite where $f_{H1} - f$ (and hence the group velocity) is smallest.

In this situation, the propagation delay as a function of frequency may be expressed as[37]

$$t = \frac{1}{c} \frac{f_{p1}f_{H1}^{1/2}}{f_{H1}'(f_{H1} - f)^{1/2}}. \tag{13.4.7}$$

The propagation delay is then given by

$$t(f) = \frac{1}{2c} \int_{\text{path}} \frac{f_{p1}(s)f_{H1}^{1/2}(s)}{(f_{H1}(s) - f)^{3/2}} ds. \tag{13.4.8}$$

Note that the proton plasma frequency $f_{p1}(s)$ and gyrofrequency $f_{H1}(s)$ both change with distance along the path, s. However, for upward propagating waves at frequencies close to the gyrofrequency, these terms vary much more slowly with distance than the term $(f_{H1}(s) - f)$. Further, as most of the propagation delay occurs very near the satellite for these waves, we may approximate the terms $f_{p1}(s)$ and $f_{H1}^{1/2}(s)$ by the values at the satellite $f_{p1}(0)$ and $f_{H1}^{1/2}(0)$ respectively. The term $f_{H1}(s) - f$ is a linear function of distance to the extent that the gyrofrequency changes linearly with distance (a very good approximation for distances of interest here). We may write, then,

$$f_{H1}(s) = f_{H1}(0) + s\, \partial f_{H1}/\partial s \tag{13.4.9}$$

$$= f_{H1}(0) + sf_{H1}', \tag{13.4.10}$$

where the prime denotes differentiation with respect to distance along the propagation path. Defining a difference frequency Δf as

$$\Delta f(s) = f_{H1}(s) - f \tag{13.4.11}$$

and using Eq. (13.4.10), we obtain

$$\Delta f(s) = f_{H1}(0) - f + sf_{H1}', \tag{13.4.12}$$

so that

$$\partial \, \Delta f / \partial s = f'_{H1}. \qquad (13.4.13)$$

Substituting (13.4.13) into Eq. (13.4.8), we find the propagation delay as a function of difference frequency from

$$t(\Delta f) = \frac{f_{p1}(0)}{2c} f_{H1}^{1/2}(0) \int_{\text{path}} \frac{\partial(s) \, d(\Delta f)}{\partial(\Delta f) \, \Delta f^{3/2}} \qquad (13.4.14)$$

$$t(\Delta f) = \frac{f_{p1}(0) f_{H1}^{1/2}(0)}{c f'_{H1}} \int_{\text{path}} \frac{d(\Delta f)}{(\Delta f)^{3/2}}. \qquad (13.4.15)$$

If we begin the integration well below the satellite where Δf is large, then

$$t(\Delta f) = \frac{f_{p1}(0) f_{H1}^{1/2}(0)}{c f'_{H1}} \left[\frac{1}{\Delta f(0)} \right]^{1/2}. \qquad (13.4.16)$$

Note that the propagation delay is dependent essentially only on the values of the parameters of the propagation medium at the satellite and the difference between the wave frequency and gyrofrequency at the satellite. This result was obtained by Gurnett and Shawhan.[37]

If we measure the wave frequency as a function of time for a proton whistler, we may plot time t against $(f^*_{H1} - f)^{-1/2}$ where f^*_{H1} is an estimated value of the proton gyrofrequency at the satellite. If the estimated value is correct, the plot will be a straight line, whereas if it is not correct the plot will be curved, with the sense of curvature dependent on whether the estimated gyrofrequency is high or low. In this way, the proton gyrofrequency (and hence magnetic field strength) at the satellite may be determined.[33]

Using a computer to determine the best fit and allowing for the change in cyclotron frequency due to motion of the satellite, the cyclotron frequency may be measured to an accuracy of 0.1%. This technique is illustrated in Fig. 25.

13.4.6. Proton Whistlers and Proton Number Density

When the proton whistler propagation delay is plotted against $(f_{H1} - f)^{-1/2}$, the slope of the straight line obtained is proportional to the proton plasma frequency, from which the proton number density is obtained. Measurements of proton number density from proton whistlers were made using this approach by Gurnett and Shawhan.[37] The estimated accuracy of this measurement is typically about 20%. Combining the fractional abundance found from the crossover frequency with the proton

---

[37] D. A. Gurnett and S. D. Shawhan, *J. Geophys. Res.* **71**, 741 (1966).

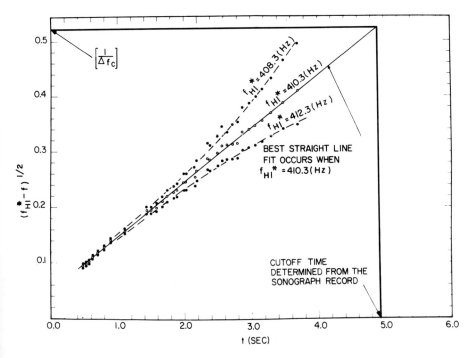

FIG. 25. A plot of t versus $(1/\Delta f)^{1/2}$ for the proton whistler (January 11, 1963; 10:59:38 UT) shown in Fig. 26. [From D. A. Gurnett and N. M. Brice, *J. Geophys. Res.* **71**, 3639 (1966).]

number density obtained from the proton whistler dispersion gives a measurement of electron density.

13.4.7. Proton Whistlers and Proton Temperature

Proton whistlers frequently show an abrupt decrease in amplitude after several seconds, as illustrated in Fig. 26. This phenomenon was investigated by Gurnett and Brice.[35] As the wave frequency approaches the proton gyrofrequency, the difference frequency Δf between the wave frequency and gyrofrequency becomes small, as does the wave phase velocity. As the doppler shift in wave frequency due to thermal motion of the proton becomes comparable to Δf, protons which "see" the wave at their cyclotron frequency absorb wave energy, causing cyclotron damping of the wave. For a Maxwellian proton velocity distribution the number of protons resonant with the wave increases exponentially as the wave frequency approaches the cyclotron frequency, giving rise to a sudden onset in damping. From the duration of the proton whistler (knowing

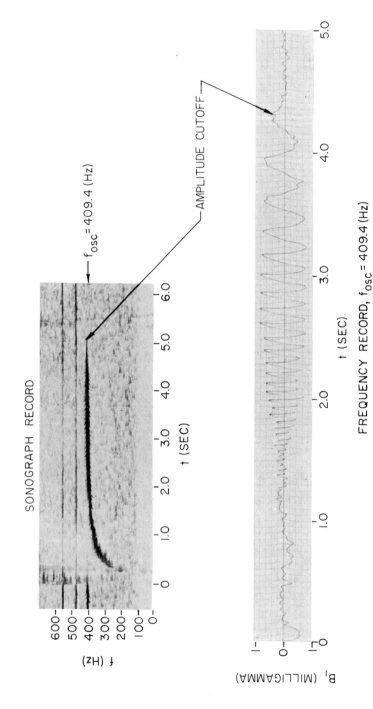

FIG. 26. Spectrogram and beat frequency oscillogram record of a proton whistler showing sharp amplitude cutoff. [From D. A. Gurnett and N. M. Brice, *J. Geophys. Res.* **71**, 3639 (1966).]

the proton plasma and cyclotron frequencies) the proton thermal velocity and hence proton temperature may be found.[35] The attenuation in nepers, β, is found to be[35]

$$\beta = \frac{\pi^{1/2}}{3c} f_{p1} \frac{f_{H1}}{f'_{H1}} \left(\frac{\Delta f}{f_{H1}}\right)^{1/2} \frac{\exp -n^3}{n^{3/2}}, \qquad (13.4.17)$$

where

$$n^3 = \frac{m_1 c^2}{2kT} \frac{f_{H1}^2}{f_{p1}^2} \left(\frac{\Delta f}{f_{H1}}\right)^3, \qquad (13.4.18)$$

where T is the proton temperature, m_1 is the proton mass, and k is Boltzmann's constant. Using Eq. (13.4.16), the attenuation as a function of time delay, t, may be put in the form

$$\beta = At^2/\exp(Bt^{-6}), \qquad (13.4.19)$$

where A and B are functions of f_{p1}, f_{H1}, f'_{H1}, and temperature T. Note that a sudden onset of damping will occur as Bt^{-6} approaches unity. The value of f'_{H1} is known from the location of the satellite, while the values of f_{p1} and f_{H1} may be found from the proton whistler dispersion as described above. The temperature is then found by measuring the time delay at which the sudden onset of damping occurs, i.e., the duration of the proton whistler.

13.4.8. Helium Whistlers

For a hydrogen–helium–oxygen mixture, there are two crossover frequencies and so both helium and proton whistlers are to be expected. Although helium whistlers occur considerably less often than proton whistlers, many fine examples of electron, proton and helium whistlers have been found by Barrington and McEwen,[38] several being shown in Fig. 27. From a knowledge of both crossover frequencies, the fractional abundances of the 3 ions may be determined. In principle the helium whistler may be used to obtain helium number density, temperature, and gyrofrequency, but helium whistlers observed to date have been relatively weak and only the crossover frequency has been measured. There is little incentive to try to determine helium number density and temperature from the helium whistler. The number density can be obtained from the proton number density (using the proton whistler) and the ion abundances (from the two crossover frequencies). Also any differences between helium and proton temperatures are expected to be very small or nonexistent.

[38] R. E. Barrington and D. J. McEwen, "Space Research," Vol. VII, p. 624. North Holland Publ., Amsterdam, 1966.

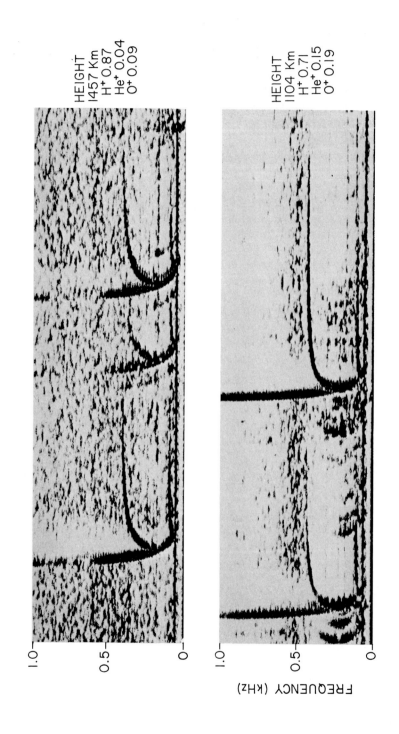

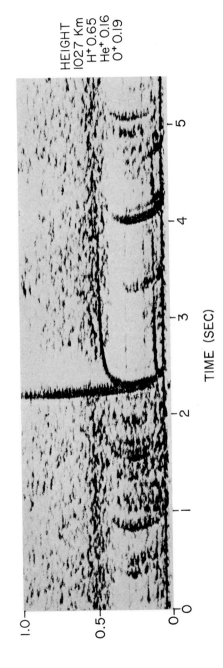

FIG. 27. Proton and helium whistlers recorded on Alouette II satellite. [From R. E. Barrington and D. J. McEwen, "Space Research," Vol. VII, p. 624. North-Holland Publ., Amsterdam, 1966.]

13.4.9. The Ion-Cutoff Whistler

The ion-cutoff whistler discovered by Muzzio[39] is an interesting phenomenon which may be considered as complementary to the proton whistler. Examples of ion-cutoff whistlers are shown in Fig. 28. The

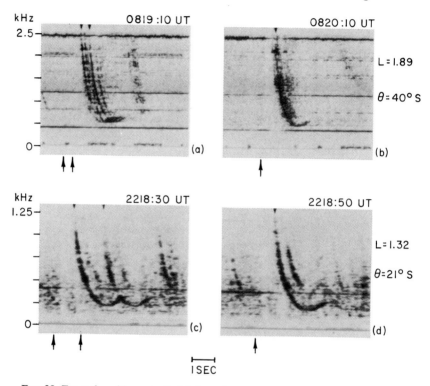

FIG. 28. Examples of ion-cutoff whistlers observed by OGO 4. (a, b) August 3, 1967; (c, d) August 5, 1967. The arrows indicate the probable time of the causative atmospheric. [From J. Muzzio, *J. Geophys. Res.* **73**, 7526 (1968).]

dispersion properties showing two time delays for a given frequency result from reflection of whistler energy at an altitude below the satellite. In the region of reflection the whistler energy propagates in the *fast* mode near the multiple-ion cutoff. This is the region of the $\omega - k$ diagram in Fig. 21 indicated by the dashed line.

The ion-cutoff whistler is a combination 1_- and 1_+ whistler which has propagated from hemisphere to hemisphere over the magnetic equator. In the hemisphere of the originating lightning flash the energy at the

[39] J. Muzzio, *J. Geophys. Res.* **73**, 7526 (1968).

lower frequencies must have coupled from the slow mode to the fast mode at the crossover frequency, which implies that the propagation direction was within the critical coupling angle. On the down-coming portion of the path, the propagation direction presumably has increased so as to be outside the critical coupling angle, or else the critical coupling angle is reduced because of different ionospheric conditions in the hemisphere of observation. As the energy propagates downward the ratio ω/Ω_p decreases, and the energy propagates with smaller and smaller values of k until it is reflected.

The minimum frequency of observation occurs very near the cutoff frequency at the point of observation. Knowing the cutoff frequency one may determine the fractional abundance of protons to an accuracy of 10%. The additional dispersion information can be used to deduce properties of the ionosphere below the satellite.

Appendix

Wave Propagation in Multiple-Ion Plasmas

For radio waves propagating in a cold, infinite, uniform, homogeneous, collisionless magnetoplasma there are two characteristic modes. For propagation parallel to the magnetic field the modes are right (R) or left (L) circularly polarized with respect to the magnetic field direction. The square of the refractive index (i.e., the relative dielectric constant) is given by

$$n^2_{\substack{R \\ L}} = 1 - \sum_r X_r/(1 \pm Y_r), \qquad (13.\text{A}.1)$$

where the summation is taken over the particle species (electrons and ions) with the X_r being the ratio of plasma frequency squared to wave frequency squared and the Y_r the ratios of gyrofrequency to wave frequency (the sign of the gyrofrequency being the same as the particle charge).

For the propagation normal to the magnetic field, the modes are denoted ordinary (o) and extraordinary (x) for which the refractive indices are given by

$$n_o{}^2 = 1 - \sum_r X_r \qquad (13.\text{A}.2)$$

and

$$n_x{}^2 = 2n_R{}^2 n_L{}^2/(n_R{}^2 + n_L{}^2) \qquad (13.\text{A}.3)$$

or, in terms of phase velocity, $W = c/n$,

$$W_x{}^2 = (W_R{}^2 + W_L{}^2)/2. \qquad (13.\text{A}.4)$$

For propagation at other angles, the phase velocity is obtained from the Astrom equation[40] and put in the following form by Smith and Brice[29]:

$$(W_R{}^2 - W^2)(W_L{}^2 - W^2)\cos^2\theta + (W_o{}^2 - W^2)(W_x{}^2 - W^2)$$
$$\times \sin^2\theta = 0. \qquad (13.\text{A}.5)$$

For a plasma with electrons and hydrogen, helium, and oxygen ions, the well-known CMA diagram has the form given in Fig. 29, which was

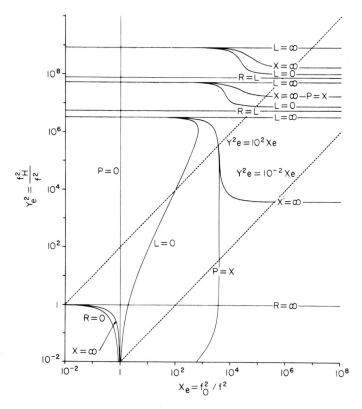

FIG. 29. CMA diagram for a multicomponent plasma. Typical ionospheric conditions lie between the two dotted lines. (From Neil Brice, "Electromagnetic Wave Theory, Part I." Pergamon Press, Oxford and New York, 1967.)

calculated for a mixture of 40% O^+ and H^+ and 20% He^+. In the figure logarithmic scales are used for X_e and $Y_e{}^2$ (the subscript "e" refers to electrons). A given plasma (constant ratio of $X_e/Y_e{}^2$) is then represented

[40] E. Astrom, *Ark. Fys.* **2**, 443 (1950).

by a 45° line in this figure. In the ionosphere and magnetosphere, the range of densities is usually such that

$$100 Y_e^2 > X_e > Y_e^2/100, \qquad (13.\text{A}.6)$$

so that the ion plasma frequencies are typically large compared to the ion cyclotron frequencies. Under this condition, all of the ion-ion cutoff ($L = 0$) frequencies and hybrid resonance ($X = \infty$) frequencies are independent of electron density and depend on ion cyclotron frequencies and relative abundances only. Crossover frequencies ($R = L$), which are not found in single-ion plasma occur at frequencies which are always independent of electron density, as indicated in Fig. 29.†

---

† ACKNOWLEDGMENT: Work at Cornell was supported in part by the Atmospheric Sciences Section of the National Science Foundation under grant NSF GA–11415.
  Work at Stanford was supported by projects under the sponsorship of agencies of the U.S. Government, including the National Science Foundation, Air Force of Scientific Research, Office of Naval Research and the National Aeronautics and Space Administration.

14. RADIO WAVE SCATTERING FROM THE IONOSPHERE†

14.1. Introduction

This part is concerned with what have been called "outdoor plasmas" rather than the indoor, laboratory variety. More specifically we will consider plasma experiments in the E region (altitudes in the vicinity of 100 km) and F region (above about 200 km) of the ionosphere.

The ionospheric plasma has several characteristics which make it attractive for certain experiments:

1. There are no important boundary effects; for many purposes the medium can really be considered to be uniform and infinite.

2. It is stationary in time as well, varying significantly only over time scales of the order of minutes or hours.

3. The ion and electron velocity distribution functions are generally very well behaved, and quasi-equilibrium approximations, possibly with different electron and ion temperatures, are almost always valid.

4. Collisions are important in some regions but can be completely neglected in others.

The outstanding disadvantage of the ionosphere from the point of view of the plasma experimenter is, of course, that there is no way to control or alter the medium. With perhaps a few exceptions, the range of parameters available for study is limited by nature.

Some typical values of various ionospheric parameters are listed in Table 1, and a typical electron density profile is shown in Fig. 1. No great accuracy is claimed for the values given in the table, but they should enable one to make quick order-of-magnitude estimates of what effects are likely to be important in a particular problem. Many of the parameters vary substantially or even drastically with time of day, season, latitude, level of solar activity, etc. The values given correspond roughly to daytime conditions at middle latitudes for moderate solar activity. At night, for example, the electron density at 100 and 300 km typically decreases 2–3 and 1 order of magnitude, respectively. More detailed descriptions of the ionosphere are available in the literature.[1]

[1] F. S. Johnson (ed.), "Satellite Environment Handbook." Stanford Univ. Press, Stanford, California, 1965.

---

† Part 14 by D. T. Farley.

TABLE 1. Typical Ionospheric Parameters for Daytime Mid-Latitude Conditions[a]

	100	300	1000	2000
Altitude (km)				
Electron density (cm^{-3})	10^5	10^6	2×10^4	10^4
Neutral particle density (cm^{-3})	10^{13}	10^9	3×10^5	2×10^4
Electron temperature (°K)	230	2×10^3	2×10^3	3×10^3
Ion temperature (°K)	230	10^3	2×10^3	3×10^3
Principal ions	NO^+, O_2^+	O^+	$O^+, H^+, (He^+)$	H^+
Plasma frequency (Hz)	3×10^6	9×10^6	1.3×10^6	9×10^5
Electron gyrofrequency (Hz)	10^6	9×10^5	7×10^5	5×10^5
Ion gyrofrequency (Hz)	18	31	—	270
Electron gyroradius (cm)	1.4	5	6	10
Ion gyroradius (cm)	4×10^2	5×10^2	—	4×10^2
Electron Debye length (cm)	0.28	0.26	2.6	3.2
Electron-ion collision frequency (sec^{-1})	10^3	6×10^2	13	4
Electron-neutral collision frequency (sec^{-1})	10^5	10	10^{-2}	10^{-3}
Ion-neutral collision frequency (sec^{-1})	5×10^3	0.6	—	3×10^{-3}
Electron thermal velocity (km/sec)	84	250	250	300
Ion thermal velocity (km/sec)	0.36	1.0	—	7.0
Electron mean free path (cm)	60	3×10^2	1.4×10^4	6×10^4
Ion mean free path (cm)	5	10^5	—	10^8

[a] Many of the quantities given vary substantially from day to night, etc. The thermal velocities are $(2KT/m)^{1/2}$.

14.1. INTRODUCTION

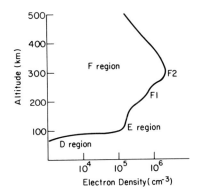

Fig. 1. A typical profile of electron density in the ionosphere.

The ionosphere has been studied for many years by a variety of radio techniques, the oldest being the ionosonde measurement in which pulses are reflected from the ionosphere and the round-trip travel time measured. By varying the transmitted frequency over a wide range up to the maximum value of plasma frequency encountered in the ionosphere (the penetration frequency), one can in principle deduce the shape of the electron density profile up to the height of the electron density maximum in the F region (see Fig. 1) but not above. The analysis of these so-called "virtual height profiles" (time delay versus frequency) to yield "true height profiles" (electron density versus altitude) is far from trivial, however (see, for example, Budden[2]). More recently, ionosonde measurements made from satellites have made it possible to study the topside of the ionosphere, the portion between the F-region maximum and the satellite.

With the advent of satellites and rockets, a variety of *in situ* measurement techniques have also come into use, providing values of local electron density and temperature, ion composition, etc. These data have advanced our understanding of ionospheric processes, and particularly their global aspects, enormously, and some experiments have revealed a number of interesting plasma resonance phenomena.

In this part we shall restrict ourselves to a discussion of radar scattering measurements made from the ground at frequencies of 50 MHz or higher, frequencies far above the maximum value of the plasma frequency (typically about 10 MHz) found in the ionosphere. At such frequencies most of the transmitted energy passes through the ionosphere and is lost; only a small fraction is scattered back to the receiver from small fluctuations in the electron density. As a result, high power transmitters

---

[2] K. G. Budden, "Radio Waves in the Ionosphere." Cambridge Univ. Press, London and New York, 1961.

and sensitive receiving equipment must be used, as well as rather sophisticated data analysis techniques. A useful introduction to a variety of ionospheric phenomena which have been studied by scatter techniques can be found in a review article by Bowles.[3] We shall explore two topics in this general area, the two which seem to have most relevance to the subject of plasma physics as a whole. The first is known as incoherent (or sometimes Thomson) scatter and involves scattering from random thermal fluctuations in the plasma density; the second is concerned with scattering from ion-acoustic waves generated by a type of two-stream instability which is encountered in certain regions of the ionosphere.

We shall begin with a discussion of some general features of all scatter measurements. Incoherent scatter and the much stronger scattering from unstable regions will then be examined in more detail. The incoherent scatter experiments, in particular, have shown remarkable quantitative agreement with theory. A very wide range of observations has been completely explained by the linearized kinetic theory. The instability theory is not so far advanced, but some aspects of the observations have been accounted for.

14.2. Scattering from a Diffuse Medium

14.2.1. Theory

Scattering from a plasma can be considered from either the single particle or continuum point of view; here we will adopt the latter approach. The operating frequency is assumed to be well above the plasma frequency and the scattering from small fluctuations in the electron density is assumed to be sufficiently weak that it does not significantly alter the transmitted wave (i.e., the Born approximation is valid). The scattering volume V_s is taken to have dimensions which on the one hand are much larger than the radar wavelength, but on the other are still small enough to allow us to treat the medium as statistically uniform. A plane electromagnetic wave incident upon this volume can be described by

$$E_{\text{inc}}(t) = E_0 \exp[i(\omega_0 t - \mathbf{k}_0 \cdot \mathbf{r})]. \tag{14.2.1}$$

The electric field of the received scattered wave can easily be shown to be[4]

$$E_s(t) = (r_e E_0 \exp[i\omega_0 t] \sin \delta / R) \int_{V_s} \Delta N(\mathbf{r}, t) \exp[-i(\mathbf{k}_0 - \mathbf{k}_s) \cdot \mathbf{r}] d^3 r, \tag{14.2.2}$$

---

[3] K. L. Bowles, *Advan. Electron. Electron Phys.* **19**, 55 (1964).
[4] F. Villars and V. F. Weisskopf, *Proc. IRE* **43**, 1232 (1955).

14.2. SCATTERING FROM A DIFFUSE MEDIUM

where R is the mean distance from the receiver to the scattering volume ($R \gg V_s^{1/3}$), ΔN is the difference between the local electron density and the mean, r_e is the classical electron radius (2.82×10^{-13} cm), $\mathbf{k}_s$ is the wave vector of the scattered wave, and δ is the angle between $\mathbf{E}_{\text{inc}}$ and $\mathbf{k}_s$.

It can be seen that (14.4.2) involves what is essentially a spatial Fourier transform of the electron density fluctuations. The scattered signal is proportional to the Fourier component with wave vector

$$\mathbf{k} = \mathbf{k}_0 - \mathbf{k}_s. \qquad (14.2.3)$$

For backscatter (transmitter and receiver at the same point, $\mathbf{k}_s = -\mathbf{k}_0$) this means the component whose planes of constant phase are perpendicular to $\mathbf{k}_0$ and whose wavelength λ is just $\lambda_0/2$ (or $4\pi/\mathbf{k}_0$). This is clearly just the spacing and orientation required to allow all the very weak partial reflections from each phase front to add coherently, since the round trip distance between two such fronts is one radar wavelength. Any other spacing results in destructive interference. For other scattering angles the appropriate wavelength becomes $\lambda_0/(2 \sin \tfrac{1}{2}\theta)$, where θ is the angle between $\mathbf{k}_0$ and $\mathbf{k}_s$. All radio scattering measurements thus involve a spatial Fourier analysis of the scattering medium. The radar is sensitive only to the spatial Fourier component whose wave vector is given by (14.2.3).

To find the scattered power and its distribution in frequency, we proceed from (14.2.2) to calculate

$$\langle E_s(\omega_0, t) E_s^*(\omega_0, t + \tau) \rangle = r_e^2 E_0^2 \sin^2 \delta \exp[-i\omega_0 \tau]/R^2$$

$$\times \iint_{V_s} \langle \Delta N(\mathbf{r}, t) \Delta N(\mathbf{r}', t + \tau) \rangle$$

$$\times \exp[-i\mathbf{k} \cdot (\mathbf{r} - \mathbf{r}')] \, d^3r \, d^3r',$$

$$(14.2.4)$$

where the angular brackets $\langle \rangle$ indicate the expected value or ensemble average. Next we make the substitution

$$\mathbf{r}' = \mathbf{r} + \mathbf{r}'' \qquad (14.2.5)$$

and note that the correlation in the density fluctuations approaches zero rapidly for very small separations (of the order of a Debye length), whereas the scale of V_s is of the order of kilometers. Therefore the integral on the right-hand side of (14.2.4) can, to a very good approximation, be replaced by

$$\int_{V_s} d^3r \int_\infty \langle \Delta N(\mathbf{r}, t) \Delta N(\mathbf{r} + \mathbf{r}'', t + \tau) \rangle \exp[i\mathbf{k} \cdot \mathbf{r}''] \, d^3r''. \qquad (14.2.6)$$

The integral in $\mathbf{r}''$ in (14.2.6) can be recognized as the spatial Fourier transform of the unnormalized space and time autocorrelation function of the electron density fluctuations. After substituting (14.2.6) into (14.2.4) it is often customary to Fourier transform both sides of (14.2.4) in time as well, giving

$$\langle |E_s(\omega_0 + \omega)|^2 \rangle \, d\omega = (r_e^2 E_0^2 V_s \sin^2 \delta / R^2) \langle |\Delta N(\mathbf{k}, \omega)|^2 \rangle \, d\omega$$

(14.2.7)

since, by the Weiner–Khintchine theorem, the Fourier transform of the autocorrelation function is just the power spectrum.

In terms of the average differential scattering cross section, (14.2.7) can be expressed as

$$\sigma(\omega_0 + \omega) \, d\omega = r_e^2 \sin^2 \delta \langle |\Delta N(\mathbf{k}, \omega)|^2 \rangle \, d\omega, \quad (14.2.8)$$

where σ is defined as the average power scattered through the angle θ (formed by $\mathbf{k}_0$ and $\mathbf{k}_s$) per unit solid angle per unit incident power per unit volume and per unit frequency interval. For backscatter measurements $\mathbf{k} \to 2\mathbf{k}_0$ and $\sin \delta \to 1$. In the literature a quantity known as the radar cross section is also often used. The latter is by definition the cross-sectional area of a perfectly conducting sphere which would scatter the same signal to the receiver, and is 4π times the cross section given in (14.2.8), a difference which at times has caused confusion.

The results given by (14.2.7) or (14.2.8) are the basic relations for scattering from a "soft" (distributed) radar target. They can as well readily be applied to phenomena such as tropospheric scatter which do not involve a plasma if $\Delta N(\mathbf{k}, \omega)$ is replaced by an appropriate expression for fluctuations in refractive index or dielectric constant. The geometry of the problem and the radar operating frequency determine the spatial characteristics of the irregularities to be observed; the time behavior of these fluctuations is then found from the power spectrum or autocorrelation function of the scattered signal. It is sometimes useful to think of the fluctuations as consisting of a sum of acoustic waves. The radar picks out a particular wavelength and orientation (a particular wave vector $\mathbf{k}$) and then determines the velocity of that wave. Of course if the waves are heavily damped, as they often are, the Doppler spectrum will be quite spread.

The above analysis also applies in principle to weak partial reflections from sharp gradients in electron density, since a step function is made up of a wide range of spatial Fourier components, one of which will be selected by the scattering process, but the physical interpretation in terms of waves is then not too meaningful. In this case the phases of the different components are not randomly distributed. For the present purposes,

14.2. SCATTERING FROM A DIFFUSE MEDIUM

however, the representation of the electron density fluctuations as the sum of a wide spectrum, in both space and time, of longitudinal, randomly phased plane waves can provide considerable physical insight into the scatter mechanism.

It can be seen from (14.2.7) and (14.2.8) that we are only interested in the statistical properties of the medium and the scattered signal. The latter is in fact noiselike; since $E_s(t)$ consists of a sum of signals from a large number of statistically independent scattering regions, it can be taken to be a Gaussian random variable whose probability density function is given by

$$p(E) \, dE = (2\pi S)^{-1/2} \exp[-E^2/2S] \, dE \qquad (14.2.9)$$

and whose joint probability density function is

$$p(E_1, E_2) \, dE_1 \, dE_2 = \frac{1}{2\pi S(1-\rho^2)^{1/2}}$$
$$\times \exp\left[-\frac{E_1^2 + E_2^2 - 2\rho E_1 E_2}{2S(1-\rho^2)}\right] dE_1 \, dE_2, \qquad (14.2.10)$$

where E_1 and E_2 are samples of E_s at times t and $t + \tau$, $S \, (= \langle E^2 \rangle)$ is the mean signal power, and ρ is the normalized autocorrelation function which depends only on τ, i.e.,

$$\rho = \langle E_s(t) E_s^*(t+\tau) \rangle / \langle |E_s(t)|^2 \rangle \qquad (14.2.11)$$

The power spectrum is the Fourier transform of ρ multiplied by S. As indicated in (14.2.11) and elsewhere, E_s is in general complex, containing both amplitude and phase information, as is ρ (but not the power spectrum, of course), but (14.2.9) and (14.2.10) can easily be suitably generalized to take account of this. Each component of E_s is Gaussian, the two components are independent, and the complex component of ρ relates the real part of E_1 to the imaginary part of E_2, etc.

The object of any scatter experiment is to estimate S and $\rho(\tau)$, or alternatively the power spectrum, as accurately as possible. This means that the signal must be averaged over some finite length of time, even if the signal strength is well above the level of the background noise.

For example, if a bank of analog filters, each of bandwidth B in hertz, is used to estimate the power spectrum, the standard deviation from the true mean of the power estimate in a particular filter, normalized by the true mean, is approximately $(2BT)^{-1/2}$, where T is the integration time; i.e.,

$$S^{-1} \left\langle [(1/T) \int_0^T |E_s(t)|^2 \, dt - S]^2 \right\rangle^{1/2} \approx (2BT)^{-1/2}, \qquad (14.2.12)$$

where S here is the expected or "true" output of the filter. The exact value of the standard deviation depends on the shape of the filter response curve. If the signal is sampled digitally, the normalized standard deviation is $M^{-1/2}$, where M is the number of independent samples; i.e.,

$$S^{-1}\left\langle [M^{-1} \sum_{i=1}^{M} |E_s(t_i)|^2 - S]^2 \right\rangle^{1/2} = M^{-1/2}. \qquad (14.2.13)$$

These last two results are equivalent, as of course they must be, since the well-known sampling theorem (see, e.g., Davenport and Root[5]) states that a band-limited signal can be completely recovered from samples taken at times separated by no more than $(2B)^{-1}$. To achieve accuracies of the order of 1%, a minimum of 10^4 independent samples is required. The same considerations apply to estimates of the autocorrelation function; 10^4 independent samples are required for each value of time delay.

When unwanted noise is introduced into the problem, the errors in the estimates of the signal parameters increase. The right-hand side of (14.2.12) and (14.2.13) must be multiplied by $(N + S)/S$, where N is the noise power, and the necessary integration times or number of samples required to achieve a particular accuracy is increased by the square of this ratio. To minimize the required integration time it is important to achieve a signal-to-noise ratio somewhat greater than unity if possible, but little is gained by increasing signal strength to much larger values.

In summary, the problems involved in scattering measurements fall into two main categories:

1. The experimenter attempts to optimize his technique so as to obtain the best possible estimate of the parameters of the incoming signal in the least time. The actual procedure used will depend on the properties of the radar, the capabilities of the available on-line computing facility, the spatial resolution desired, etc.

2. The theorist endeavors to interpret the experimental results in terms of the parameters (density, temperature, drift velocity, instability level, etc.) of the plasma under investigation.

In the rest of this part we shall see to what extent these problems have been solved in two particular classes of scatter measurements.

14.2.2. Measurement Techniques

Radar studies of the ionosphere have evolved through the years, and the techniques, particularly for incoherent scatter measurements, have

[5] W. B. Davenport, Jr., and W. L. Root, "An Introduction to the Theory of Random Signals and Noise." McGraw-Hill, New York, 1958.

become rather sophisticated. No attempt will be made here to discuss the whole field of radar and communications technology, but we will try to point out some of the more important or perhaps less widely appreciated features of the techniques. We shall not be concerned with the design of high power transmitters, low noise receivers, sampling equipment, etc.

The three features of the scattered signal that can be measured are the total power, the polarization, and the spectral characteristics. We shall discuss the usual methods used for these measurements as well as a few more sophisticated techniques which are useful in some situations. Further details on several of these topics may be found in a review by Evans[6] of incoherent scatter.

14.2.2.1. Power. The power is the simplest of the signal parameters, but even this measurement may require considerable care when the signal is much weaker than the background noise, as it often is in the case of incoherent scatter. It is generally much easier to determine the relative scattering cross section as a function of time or range than the absolute cross section. It is often difficult to calibrate the transmitter power, antenna efficiency, line losses, receiver gain, etc. with sufficient accuracy to make absolute measurements. It is possible to hold these quantities fairly constant, however, thereby making relative measurements possible. Even this requires considerable care, though, as anyone who has worked with high-power radars can testify. When the transmitter and receiver share the same antenna, receiver recovery can be a problem; the receiver is shut off during the transmitter pulse, and some time is required for it to settle after it is switched on again. Moreover various components, particularly switches, may be heated during the pulse and may cause the noise level of the receiver to vary as a function of pulse length and time after the pulse. In cases where the dynamic range of the received signal is large, the receiver detector response (linear, square-law, etc.) may not be uniform over the whole range.

These are but a few of the problems that have been met, and for the most part solved, in scattering measurements. They are certainly not trivial problems, but they will not be discussed further here.

Since unwanted background noise is always present in the system, it is generally necessary to determine the signal power from separate measurements of the signal plus noise and the noise alone. The latter is found by either repeating the first measurement with the transmitter turned off, or by measuring the noise simultaneously at a neighboring frequency or orthogonal polarization where there is no signal. One must be sure that

[6] J. V. Evans, *Proc. IEEE* **57**, 496 (1969).

the noise level is not changed by the presence of the signal or by the change in frequency or polarization. In some cases any differences that do exist may be compensated for by interchanging the signal and noise channels alternately from one frequency or polarization to the other. As a general rule of thumb, it is always safest to assume that any systematic error that could be present *will* be present, at least at times, in spite of valiant efforts to eliminate it. If at all possible, a switching (preferably digital) scheme should be devised which will cause practically all of the residual errors to cancel out. Such schemes are an integral part of all incoherent scatter measurements (not just of power). In other scatter studies, the required accuracy is often not as great and the systematic errors are not as important.

Statistical and computational considerations are sometimes important in determining how best to process the incoming data. The incoming signal and noise are usually both Gaussian random variables whose variance (mean power) we wish to estimate as accurately as possible. An obvious method would be to square-law detect the signal and then average the result, using either analog or digital integration, or some hybrid of the two. In practice, analog integration turns out to be unsatisfactory in demanding applications such as incoherent scatter. Completely digital processing is usually used, although hybrid techniques may yet prove useful in some applications.

Although square-law detection is in fact optimum from a statistical point of view, it is seldom used in incoherent scattering. The detectors are usually not sufficiently accurate over a large dynamic range of the input signal, and the large dynamic range of the output caused by the squaring process may be inconvenient for analog-to-digital conversion and subsequent data processing. The many bits required for the data words increase the cost and/or decrease the speed of the processing.

These problems can be avoided in a number of ways.[7] The optimum method depends upon the available computing facilities. In some instances, for example, it may prove convenient to digitize the IF output of the receiver and determine the power by summing (averaging) the absolute value of this quantity. This is rather a departure from more conventional methods utilizing square-law or linear envelope detection, but it is only slightly less efficient, statistically. The loss may be more than compensated for by a reduction in systematic errors and/or an improvement in data processing speed. Most incoherent scatter measurements, at least, are computer limited; more data are available than can be processed in the time between transmitter pulses.

[7] D. T. Farley, *Radio Sci.* **4**, 139 (1969).

14.2.2.2. Polarization. Radar measurements can be used to study the differential Faraday rotation of a radio wave passing through the ionosphere, and hence the electron density profile. The measurement is in principle very straightforward. As is well known, it can be easily shown from the equations of magneto-ionic theory[8] that the plane of polarization of a linearly polarized plane wave of suitable frequency will rotate as it passes through a plasma in a direction parallel to the magnetic field. This happens because the two characteristic modes of propagation in this case are circularly polarized, with slightly different refractive indices. In the ionosphere, and for typical radar frequencies, these results are approximately true (quasilongitudinal approximation) unless the propagation direction is almost precisely normal to the magnetic field.

If ϕ is defined as the angle of the plane of polarization, the electron density can be found from

$$N(h) = K(h)\, d\phi/dh, \qquad (14.2.14)$$

where K is a factor proportional to $f^{-2}B \cos \alpha$, α is the angle between **k** and **B**, and f is the operating frequency. The factor K can be computed from magnetic field models. By measuring ϕ as a function of range, $N(h)$ can be easily found. In contrast to the usual Faraday rotation studies of the ionosphere, in which satellite transmissions are used and only the integrated total electron content weighted by the magnetic field factor is found, the radar measurements allow the complete shape of the profile to be determined. Furthermore, one obtains absolute electron densities which are unaffected by variations in transmitter power, etc.

While the measurement is simple in principle, in practice it is fraught with considerably more difficulties than the power measurements. A variety of rather subtle systematic errors can seriously distort the results. These errors and their remedies are discussed in detail elsewhere.[9] One of the important sources of error is the depolarized signal caused by the Faraday dispersion due to the variation of $\cos \alpha$ across the beam. The elimination of this error generally requires the use of some sort of cross correlation or sample-by-sample multiplication of the phase coherent signals received on two orthogonal polarizations (which could be circular as well as linear). Most other systematic errors arising from equipmental imperfections such as gain imbalances, phase errors, biases, etc. can either be made to cancel through the use of a variety of switching techniques, or else they can be accurately measured and taken account of in subsequent analysis of the data.

[8] J. A. Ratcliffe, "The Magneto-Ionic Theory and its Applications to the Ionosphere." Cambridge Univ. Press, London and New York, 1959.

[9] D. T. Farley, *Radio Sci.* **4,** 143 (1969).

14.2.2.3. The Spectrum and Autocorrelation Function. In some experiments it is satisfactory and convenient to measure the spectrum of the signal directly, using a bank of filters; often, however, better methods are available. When using filter banks, it is necessary only to ensure that the filters have reasonable cutoff characteristics and that any associated amplifiers have properly calibrated gains. Periodic calibrations can be carried out by feeding white noise into the entire bank, for example.

The spectrum of the scattered signal will be simply the convolution of the spectrum of the scattering cross section with that of the transmitted signal. Thus the resolution with which we can measure the spectral characteristics of the scattering cross section is of the order of the inverse of the duration of the transmitted pulse. With an ordinary radar having a single antenna, the range resolution is of the order of the pulse length. Often the two resolution requirements are incompatible. We would like to use short pulses for range resolution, but still manage to determine the Doppler spectrum of the scattering cross section.

Two approaches are possible. One option is to avoid the difficulty completely by making a bistatic measurement, using separate antennas for transmission and reception. The range resolution is then determined by the intersection of the two beams, and so arbitrarily long pulses may be used. On the other hand, only one region of space (or perhaps several if multiple beams are used) can be studied at a time. Further discussion of the use of the bistatic technique in incoherent scatter is given by Evans.[6]

The second possibility is to measure the autocorrelation function (14.2.11) of the scattered signal, i.e., the Fourier transform of the power spectrum. This topic is discussed in detail by Farley.[10] In many situations the correlation technique makes it possible to "have your cake and eat it too"; adequate resolution in both range and frequency (or equivalently, time delay) often can be obtained, even though only a single antenna is used (monostatic operation).

Three possible procedures for measuring the correlation function when the medium under study is distributed over a considerable range of altitude (as in incoherent scatter studies) are illustrated in the height–time diagrams shown in Fig. 2. In the first method shown, two pulses having the same length but orthogonal polarizations are transmitted at times 0 and τ. The two polarizations correspond to the two magneto-ionic modes of the plasma and are usually right and left circular. Two separate receivers are used, and the appropriate receiver outputs are sampled at times $2h/c$ and $2h/c + \tau$. Signals from a particular pulse enter only one

[10] D. T. Farley, *Radio Sci.* **4**, 935 (1969).

14.2. SCATTERING FROM A DIFFUSE MEDIUM

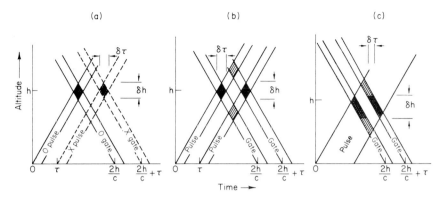

FIG. 2. Schematic range-time diagrams illustrating three possible techniques for measuring the temporal autocorrelation function of the scattering cross section. In (a) the two transmitted pulses have orthogonal polarizations; in (b) and (c) only one polarization is used. The regions from which signals enter the appropriate receiver gates are shaded, and the approximate resolution in range and time delay is indicated. [From D. T. Farley, *Radio Sci.* **4**, 935 (1969).]

receiver. The altitude range from which information is received at a given time by a particular receiver is shown shaded on the diagram. The receiver gate has a finite width in time because of the finite receiver bandwidth.

The two outputs are now digitized and multiplied together. From an average of the results of many pulse pair transmissions, we obtain an estimate of $\langle E_s(h, t)E_s^*(h, t + \tau)\rangle$, the true expected value or ensemble average of the cross product. In practice, of course, a great many altitudes can be sampled after each pulse pair transmission. By cyclicly changing the spacing between the pulses and the corresponding receiver samples, a series of time lags τ in the correlation function can be studied. With some radar systems it is possible to have the pulses overlap (τ less than the pulse length) or be coincident, but usually this cannot be done. The cross products for each altitude and time delay are summed separately in a computer, and finally the results are normallized in some suitable way. Different normalization procedures may give somewhat different values for the mean-square statistical errors.[10]

Cosmic noise and receiver noise, as well as signal, will be present in each output channel, but since the two polarizations are orthogonal and separate receivers are used, the two noise outputs are statistically independent of each other and of the scattered signal. Thus terms involving noise make no net contribution to the correlation function estimate; they contribute only to the statistical fluctuations.

In the second double-pulse method shown, the two pulses have the same polarization and only one receiver channel is used. As a result, signals from two altitude ranges (shown shaded) are contained in each sample. However, since signals from different altitudes are uncorrelated, only those from altitudes in the vicinity of h (the heavily shaded regions) contribute to the mean of the cross products. Signals (clutter) from the other two altitude ranges (lightly shaded) contribute to the noise. The maximum signal-to-noise ratio obtainable with this technique is roughly unity, whereas in the previous case it could be made arbitrarily large. Furthermore, the cosmic noise and receiver noise terms will now in general contribute to the mean of the cross product, so a separate estimate of these must be made and subtracted from the result for signal plus noise, in order to obtain the correlation function of the signal alone.

The third method (Fig. 2c) is similar to the second, except that in this case only a single long pulse is transmitted. Again the two samples indicated contain signals from the shaded range of altitude and time, but only contributions from the darkly shaded portions are correlated. In contrast to the other two methods, here both the strength of the correlated signal and the mean altitude to which it corresponds depend upon the pulse length and time lag τ, as well as the time of the first sample.

Only a single time lag is indicated on the single pulse diagram, but it is possible and desirable to measure a number of time delays simultaneously, taking account in the analysis of the fact that the different lags correspond to slightly different common volumes and therefore signal strengths. If the pulse length is considerably longer than the longest time delay one wishes to examine, it is possible to measure all time delays at all altitudes after each pulse transmission. Unfortunately, it is often not possible to make such full use of all the available information because of computer limitations.

The resolution in altitude and time lag obtainable with the different techniques is indicated in a rough schematic way in Fig. 2. A more complete discussion has been given by Farley.[10] The true variation of $\rho(h, \tau)$ with τ is convolved in the measurement with a function whose width in time is approximately the pulse width or the gate width (reciprocal of the receiver bandwidth), whichever is *smaller*. In the single-pulse method, the gate width is always the determining factor. With the double-pulse methods, the true variation of $\rho(h, \tau)$ with altitude is convolved with a function whose width is roughly one half the velocity of light multiplied by the pulse duration or gate width, whichever is *larger*. In the single pulse measurement, the height resolution depends on both the pulse length and the time lag, as can be seen from Fig. 2c.

Various modifications and hybrid combinations of the basic procedures just discussed are possible and have been used. For example, measurements have been made in which four pulses, rather than two, were transmitted in an unequally spaced group. This mode of operation makes it possible to sample six different time delays simultaneously, while maintaining the altitude resolution of a single pulse. The main disadvantage of this technique is that a considerable amount of unwanted signal (clutter) is added to the noise.

Most of the discussion of this section has been particularly relevant to incoherent scattering measurements, in which the extent of the region being investigated is much larger than a typical pulse length and corresponds to time intervals much longer than the time lags of interest in the correlation function (i.e., $2 \Delta h/c \gg$ typical τ). In radar studies of the equatorial electrojet and the aurora, however, this is often not the case; the "target" is confined in range and the time delays of interest are comparatively long. In such cases the correlation function can be measured in a straightforward way. The transmitter is pulsed periodically at a fairly rapid rate, but still slow enough to ensure that signal returns from different pulses do not overlap, in contrast to the incoherent scatter measurements (Fig. 2). Such a situation is sketched in Fig. 3. The signal is sampled at a

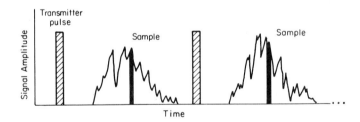

FIG. 3. A sketch of the signal returned from a limited scattering region, such as the equatorial electrojet. In this case the time taken for the medium to change significantly is much longer than the time between pulses.

particular range and digitized, and the result is multiplied by corresponding results for many preceding pulses. As many as 32 or 64 time delays are studied simultaneously in a typical measurement. If computing capability permits, more than one range at a time can also be sampled. Another possibility often resorted to is to tape record the incoming signal and play it back a number of times, each time measuring the correlation function at a different range.

14.2.2.4. The One-Bit and Hybrid Autocorrelation Functions. We have referred before to the fact that many scatter studies of the ionosphere are computer limited. The incoming data cannot be fully processed by the on-line digital equipment in the time available, and as a result a substantial portion of the data (sometimes more than 99%!) must be thrown away. This is particularly true of correlation measurements, in which a large number of time-consuming multiplications and additions must be performed. The digital processing obviously can be greatly simplified and speeded up if only the algebraic sign of the incoming signal voltage is used, but at first glance it might seem that hardly any information would be left if only the sign were retained. Surprisingly, however, this is not the case. For noiselike signals (Gaussian random variables) at least, the complete correlation function can be recovered from the hard limited or infinitely clipped version of the signal.

If we define $x(t)$ to be the sign of the incoming signal, then the one-bit correlation function $r(\tau)$ is given by

$$r(\tau) = \langle x(t)x(t+\tau) \rangle. \tag{14.2.15}$$

Van Vleck and Middleton[11] have shown that $\rho(\tau)$, the correlation function of the undistorted signal, and the one-bit correlation function are related by

$$\rho(\tau) = \sin[\tfrac{1}{2}\pi r(\tau)]. \tag{14.2.16}$$

For the same number of independent samples, the statistical errors are larger in the one-bit case than when the complete signal is used. The standard deviation from the expected or "true" value is larger by a factor[12] of about $\pi/2$, or in other words about $\pi^2/4$ times as many independent samples must be used in the one-bit case to achieve the same statistical accuracy (since the standard deviation is proportional to the inverse of the square root of the number of independent samples). When the samples are not independent, the difference in number of required samples becomes somewhat smaller.

We see that something like twice as many multiplications must be performed in the one-bit case to get results comparable to those obtained using multibit techniques. This is a small price to pay, however, if the data-taking rate can be increased by two or more orders of magnitude. One additional drawback of the one-bit technique is that all information about the total power of the received signal is lost. This is seldom a serious problem; the power can be measured separately.

A hybrid technique, in which only the sign of all the past samples is retained, but the present sample is fully digitized, can also be used. This gives up some of the simplicity of the fully one-bit technique, but recovers some of the lost information. The correlation function derived from these

---

[11] J. H. Van Vleck and D. Middleton, *Proc. IEEE* **54**, 2 (1966).
[12] S. Weinreb, Tech. Rep. 412. Res. Lab. of Electron., M.I.T., Cambridge, Massachusetts, 1963.

cross products will be undistorted; no correction formula such as (14.2.16) need be applied. The power in the signal can also be found.

The multiplications in the hybrid techniques involve only the sign bits, and so are just as simple as in the one-bit case. The summations, which could be done with a simple counter in the one-bit case, are now more cumbersome, however.

Further discussion of the one-bit and hybrid techniques and their application to radar studies of the ionosphere is given elsewhere.[10] It turns out, as might be expected, that in terms of statistical errors, the hybrid method falls midway between the one-bit and multibit methods. Approximately $\pi/2$ times as many independent samples as in the full multibit case are required to achieve the same statistical accuracy.

14.2.2.5. Higher-Order Spectral Functions. Virtually all spectral analysis of radar signals scattered from the ionosphere has been concerned with the power spectrum or autocorrelation function. Higher order spectral functions can also be calculated, however, and may prove to be of some interest. The simplest of these is the bispectrum. If, for simplicity, we take $f(t)$ to be a real stationary random function of time, the bispectrum $B(\omega_1, \omega_2)$ can be defined as

$$B(\omega_1, \omega_2) = (1/(2\pi))^2 \int\!\!\int_{-\infty}^{\infty} \langle f(t)f(t + \tau_1)f(t + \tau_2)\rangle$$
$$\times \exp(-i\omega_1\tau_1 - i\omega_2\tau_2)\, d\tau_1\, d\tau_2. \quad (14.2.17)$$

It is essentially the two-dimensional Fourier transform of the second-order correlation function, the term inside the angular brackets denoting ensemble average. For a truly Gaussian process this correlation function and the bispectrum vanish. They do not vanish, on the other hand, if nonlinear processes are important and the different Fourier components of the signal spectrum are not statistically independent.

Thus these functions, and perhaps others of still higher order, may be of use in detecting the presence of nonlinear processes and studying their properties. The bispectrum, for example, has been used in the analysis of the interaction of ocean waves in shallow water.[13] Bowles[13a] has suggested a number of possible situations in which these higher-order functions might be of interest in radar investigations and for which the interpretation is relatively simple. One example arises if the scattering is

---

[13] K. Hasselmann, W. Munk, and G. MacDonald, *in* "Time Series Analysis" (M. Rosenblatt, ed.), Chapter 8. Wiley, New York, 1963.
[13a] K. L. Bowles, Private communication, 1969.

mainly from a small number of fairly discrete, drifting, amplitude modulated irregularities; i.e., if $f(t)$ is roughly of the form

$$f(t) = \cos(\omega_0 t + \phi_0)[1 + A_m \cos(\omega_m t + \phi_m)] \qquad (14.2.18)$$

with $\omega_m \ll \omega_0$ and $A_m \lesssim 1$.

The higher-order spectra may well prove to be useful in the study of scattering from the unstable plasma waves found in the equatorial electrojet and the aurora since here nonlinear effects are certainly significant. One problem with these functions is that far more computer operations are required to calculate them than are needed for the usual spectral and correlation function analysis.

14.2.2.6. *Frequency Stepping.* In any radar measurement there is a maximum possible pulse repetition frequency (PRF), which may be determined by the average power capacity of the transmitter, the on-line computer capability, etc. In some cases it may be determined by confusion; sufficient time between pulses must be allowed so that significant amounts of signal power are not received from two altitude regions at the same time. This confusion can be avoided by changing the transmitter and receiver frequency from one pulse to the next by an amount greater than the receiver bandwidth. Sampling rates in certain measurements have been increased by as much as a factor of ten (ten different frequencies were used) through the use of this procedure.

14.2.2.7. *Pulse Compression.* Pulse compression is a close relative of frequency stepping. The latter is essentially an incoherent technique whose object is to increase the average transmitter power; the former is a coherent technique which in effect increases the peak power. Pulse compression (frequency "chirping" or phase coding) is frequently employed in radar detection work,[14] but has only recently been considered for ionospheric diagnostic measurements, where the target is diffuse and introduces a wide spectrum of Doppler shifts.

The basic idea of the technique is that in the course of transmission of a relatively long pulse, the transmitter frequency and/or phase are varied in such a way that a wide frequency spectrum with known phase relationships is transmitted. The frequency band is just that of a very short pulse, but the phases are altered. Since the phases are known, however, the received signal can be processed in such a way that it very closely resembles the signal that would have been received if a very short pulse, with the total energy of the long pulse, had in fact been transmitted. Hence the term pulse compression. It is particularly useful if good range resolution is needed in a region of low signal-to-noise ratio.

[14] C. E. Cook and M. Bernfeld, "Radar Signals." Academic Press, New York, 1967.

Pulse compression can be achieved in various ways; here we will just briefly mention one of them—known as Barker coding. In this technique the phase within the transmitted pulse is switched by 180° several times. The returning signal is "decoded" by cross-correlating it with a replica of the transmitted pulse. As an example, a five-bit Barker code and its autocorrelation function are shown in Fig. 4. In (a) the transmitted

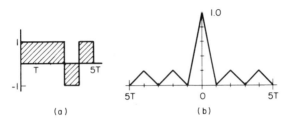

FIG. 4. A five-bit Barker code and its autocorrelation function. In (a) the 180° phase change is indicated by −1. The autocorrelation function (b) has $n − 1$ "sidelobes" with amplitude n^{-1}.

pulse is shown, with +1 and −1 corresponding to phases of 0° and 180°. The autocorrelation function (b) can be thought of as the signal voltage, after decoding, corresponding to an echo from a small stationary discrete target. A single uncoded pulse of length T would give only the central spike.

Since the signal voltages add coherently, while the noise adds incoherently, the signal-to-noise ratio for the central spike is improved by a factor of n for an n-bit code relative to the value for a single pulse of length T (one bit) of the same peak power. However, $n−1$ "side-lobes" or clutter echoes are also introduced. The ratio of the power in each clutter echo to that of the central echo is n^{-2}; hence the overall signal-to-clutter ratio is $n^2/(n − 1)$. The maximum possible value of n for which a Barker code exists is 13. With longer codes the "sidelobes" are not uniform and some have relative powers greater than n^{-2}.

When the target introduces Doppler shifts into the received signal, as is always the case in scatter measurements, the analysis becomes more complex and a quantity known as the ambiguity function must be considered. This function is dealt with in detail by Cook and Bernfeld.[14] For our purposes the simple description given here is often nearly correct, so long as the correlation time of the medium is at least as long as the full transmitted pulse. Coherent integration, which is the basis of any pulse compression scheme, is obviously not possible if the medium itself loses coherence in the course of passage of the pulse.

The phase-coded signal can be decoded with an on-line general purpose computer, but in some applications this procedure will be too slow. It is better to decode the signal immediately after analog-to-digital conversion with special purpose digital hardware. If the code is long, one-bit correlation techniques may be useful.

14.3. Incoherent Scattering

14.3.1. Introduction

Gordon[15] was the first to point out that it should be possible to detect scatter from individual electrons in the ionosphere. If the electrons are randomly distributed in space, the phases of the individual contributions to the total scattered signal will also be randomly distributed, and the total signal strength can be found by simply summing the powers of the separate scattered waves. In contrast to this incoherent scatter, for coherent or partially coherent scatter from meteor trails, unstable plasma waves, etc., the phases of the individual scattered waves are at least partially correlated, and it is essentially the voltages rather than the powers which must be summed. In this latter case the signal will generally be quite strong and easily detected; in the former, as we shall see, the signal is very weak.

The theory of scattering from a single free electron was first worked out by Thomson[15a] (hence the alternate name Thomson scatter, which is sometimes used in the ionospheric literature), the discoverer of the electron. The *radar* scattering cross section for a single electron is

$$\sigma_e = 4\pi r_e^2 \sin^2 \delta \approx 10^{-24} \sin^2 \delta \quad \text{cm}^2. \tag{14.3.1}$$

To get a rough appreciation of the order of magnitude of the radar detection problem, consider the total scattering cross section corresponding to a typical scattering volume at an altitude of about 300 km, near the altitude of maximum electron density in the ionosphere. For a scattering volume measuring 5 × 5 × 10 km say, corresponding roughly to a backscatter measurement with a 1° antenna beamwidth and a 67 μsec pulse length, with a mean electron density of $10^6/\text{cm}^3$, the total radar cross section is about 0.25 cm². This would seem at first glance to be a very small target indeed at a range of 300 km. Gordon[15] pointed out, however, that modern (in 1958) radars could in fact detect such a tiny target, even taking into account the fact that the bandwidth of the scattered signal was

[15] W. E. Gordon, *Proc. IRE* **46**, 1824 (1958).

[15a] J. J. Thomson, "Conduction of Electricity through Gases." Cambridge Univ. Press, London, 1906.

expected to be quite large, because of the thermal motion of the electrons. To be sure, the required transmitter powers were measured in megawatts and the antenna areas in acres, but everything necessary was within the state of the art.

Soon after Gordon's prediction, Bowles[16] observed the effect in the ionosphere, using an existing transmitter at 41 MHz (not an optimum frequency) with a power of 4–6 MW and a simple but large vertically directed antenna which was constructed in a few weeks and consisted of 1024 half-wave dipoles. Bowles' first measurements were quite crude, but they showed clearly that the simple scatter theory proposed by Gordon was incorrect; the observed bandwidth of the scattered signal was much narrower than that predicted on the basis of completely free electrons. The total scattered power was roughly that expected, however.

It was quickly realized that the ions play an important role in the scattering process, even though it is of course the electrons that do the scattering. Only when the radar wavelength is much smaller than the Debye length (which is usually *not* the case in practice in the ionosphere; see Table 1) can the electrons be treated as completely independent particles. The general theory for large or small radar wavelengths was soon worked out by a number of authors using a variety of techniques. The results are happily nearly all in agreement, and it is now possible to predict quantitatively the total power and the power spectrum of the scattered signal for almost any conceivable set of ionospheric parameters.

There were many early difficulties with the measurements. It was found to be one thing to detect the effect, quite another to measure it accurately. The very weak signals often require long integration times and are subject to a variety of subtle systematic errors. With improved equipment and the use of sophisticated data processing techniques such as those we have just examined, however, these problems have been largely overcome. There are now a number of major observatories devoting a substantial amount of time and effort to incoherent scatter observations.

The two largest facilities are shown in Figs. 5 and 6. The Jicamarca Radio Observatory is located near Lima, Peru and is almost on the magnetic equator. The radar operates at a frequency of 50 MHz with a peak power of about 4 MW. The antenna consists of a square array of 9216 pairs of crossed dipoles covering an area of about 8×10^4 meters^2. The Arecibo Observatory in Puerto Rico has a radar which operates at a frequency of 430 MHz with a peak power of about 2 MW. The antenna has been placed in a natural limestone sink hole, and the reflector is a section of a sphere with diameter of about 300 meters. The

---

[16] K. L. Bowles, *Phys. Rev. Lett.* **1**, 454 (1958).

FIG. 5. A view of the Jicamarca Radio Observatory, located near Lima, Peru. The antenna consists of an array of 18,432 half-wave dipoles (courtesy of ESSA Research Laboratories).

beam is steered by moving the line feed which is suspended above the reflector surface.

A comprehensive review of current theory and practice in incoherent scattering measurements is given in the previously referred to work of Evans.[6] He also describes many of the important experimental results which have been obtained. A shorter summary of the current status and future potential of the incoherent scatter technique is given in a review by Farley.[17]

14.3.2. Theory

14.3.2.1. Method of Derivation. The problem of calculating the spectrum $\langle | \Delta N(\mathbf{k}, \omega) |\rangle^2$ of the random fluctuations of electron density in a plasma, and thereby finding the scattering cross section using (14.2.8) has been tackled by a considerable number of authors, only a sampling of whom will be referred to in this brief account. The total scattered

[17] D. T. Farley, *J. Atmos. Terr. Phys.* **32**, 693 (1970).

14.3. INCOHERENT SCATTERING

FIG. 6. A view of the Arecibo Observatory. The antenna reflector surface is 1000 ft in diameter and is set in a natural limestone sinkhole near Arecibo, Puerto Rico.

power, but not the frequency spectrum, can in some cases be found from the Debye–Huckel theory.[18,19] For a collision-dominated plasma, the spectrum can be found using the fluid equations[20,21] which can also provide a rough qualitative understanding of the scattering even in the collisionless case.[22] Most of the theoretical work has been based on the kinetic equations, however.[19,23–30]

[18] E. E. Salpeter, *Phys. Rev.* **120** (1960).
[19] J. Renau, *J. Geophys. Res.* **65**, 3631 (1960).
[20] B. S. Tanenbaum, *Phys. Rev.* **171**, 215 (1968).
[21] R. G. Seasholtz and B. S. Tanenbaum, *J. Geophys. Res.* **74**, 2271 (1969).
[22] M. H. Cohen, *J. Geophys. Res.* **68**, 5675 (1963).
[23] J. A. Fejer, *Can. J. Phys.* **38**, 1114 (1960); **39**, 716 (1961).
[24] E. E. Salpeter, *Phys. Rev.* **122**, 1663 (1961).
[25] T. Hagfors, *J. Geophys. Res.* **66**, 1699 (1961).
[26] J. P. Dougherty and D. T. Farley, *Proc. Roy. Soc. Ser. A* **259**, 79 (1960).
[27] J. P. Dougherty and D. T. Farley, *J. Geophys. Res.* **68**, 5473 (1963).
[28] D. T. Farley, J. P. Dougherty, and D. W. Barron, *Proc. Roy. Soc. Ser. A* **263**, 238 (1961).
[29] D. T. Farley, *J. Geophys. Res.* **71**, 4091 (1966).
[30] M. N. Rosenbluth and N. Rostoker, *Phys. Fluids* **5**, 776 (1962).

The work of the present author has involved the use of the fluctuation-dissipation or Nyquist theorem, whereby random thermal fluctuations in some particular quantity (such as the electron density) can be related to an appropriate dissipation process (such as Landau damping or collisions). The Nyquist theorem applies rigorously only to situations involving complete thermal equilibrium, but it can be generalized to quasi-equilibrium situations in which, for example, the electron and ion temperatures are not equal, just as one can easily calculate the noise output of an electrical circuit containing impedances at two different temperatures.

All of the kinetic theory approaches to the problem lead to results for the differential scattering cross section (14.2.8) which can be expressed as

$$\sigma(\omega_0 + \omega)\, d\omega = N r_e^2 \Gamma \sin^2 \delta\, d\omega/\pi\omega, \qquad (14.3.2)$$

where

$$\Gamma = \frac{|y_e|^2 \operatorname{Re}(y_i) + |\mu y_i + ik^2 \lambda_D^2|^2 \operatorname{Re}(y_e)}{|y_e + \mu y_i + ik^2 \lambda_D^2|^2}, \qquad (14.3.3)$$

μ is the temperature ratio T_e/T_i, λ_D is the electron Debye length, and Re denotes the real part of a complex quantity. The terms y_e and y_i are normalized complex electron and ion admittance functions (closely related to the appropriate mobility or conductivity), for electric fields proportional to $\exp[i(\omega t - \mathbf{k}\cdot\mathbf{r})]$. For a plasma containing a magnetic field, the response of the particles to an electric field is of course described by a tensor, but only the "longitudinal" term, the term which describes the response in the direction of the applied electric field, is relevant to the scattering process. This result for the scattering cross section can easily be generalized to include the effect of ion mixtures or multiply charged ions.[28]

The functions y_e and y_i can be found from the Vlasov equation in the usual manner when collisions can be neglected, which is often but not always possible in the ionosphere. If we include the effect of a uniform magnetic field, the result, for both the ions and electrons, can be expressed as

$$y = 1 + \theta \int_0^\infty \exp[-i\theta t' - \phi^{-2} \sin^2\alpha \sin^2 \tfrac{1}{2}\phi t' - \tfrac{1}{4} t'^2 \cos^2\alpha]\, dt', \qquad (14.3.4)$$

where α is the angle between $\mathbf{k}$ and the magnetic field $\mathbf{B}$, t' is dimensionless, and θ and ϕ are the normalized Doppler shift and Larmor frequencies; i.e.,

$$\theta = (\omega/k)(m/2KT)^{1/2}, \qquad \phi = (\Omega/k)(m/2KT)^{1/2}, \qquad (14.3.5)$$

where $\Omega = eB/m$ in MKS units, K is Boltzmann's constant, and the mass and temperature appropriate to the particle species must be used. The

14.3. INCOHERENT SCATTERING

integral in (14.3.4) is the Gordeyev integral. The normalized frequency θ can also be thought of as the ratio of the wave velocity to the thermal velocity, whereas ϕ is essentially the scale size k^{-1} divided by the Larmor radius.

In practice the general relation (14.3.4) for y can often be approximated by simpler expressions that are more convenient for numerical calculations. Asymptotic expansions are usually possible when ω is large (e.g., $\omega \sim \omega_p$, the plasma frequency). For small Doppler shifts ($\theta \lesssim 1$) the magnetic field may be sufficiently weak ($\cos^2 \alpha \gg \phi^2$), as it usually is for the ions in the ionosphere, or strong ($\phi^2 \gg 1$), as it often is for the electrons, to permit $y(\theta)$ to be expressed in terms of the plasma dispersion function, which can be calculated easily using methods discussed by Fried and Conte.[31] When $\alpha \to \pi/2$, electron or ion cyclotron frequency effects may become important, although the ion effects are often destroyed by ion-ion collisions.

In the lower part of the ionosphere, collisions between the charged and neutral particles (especially ion-neutral collisions) become important. The Boltzmann equation for the distribution function $f(\mathbf{x}, \mathbf{v}, t)$, from which the admittances are calculated, can be written as

$$(\partial f/\partial t) + \mathbf{v} \cdot \nabla f + (1/m)q(\mathbf{E} + \mathbf{v} \times \mathbf{B}) \cdot \nabla_v f = (\partial f/\partial t)_c \quad (14.3.6)$$

in the usual notation where $q = \pm e$ and the term on the right-hand side is the collision term. In many problems of this sort the collision term can be approximated by a simple relaxation expression; i.e.,

$$(\partial f/\partial t)_c = -\nu(f - f_0) = -\nu f_1, \quad (14.3.7)$$

where ν is some effective collision frequency and f and f_0 are the perturbed and unperturbed distribution functions. In this case the only alteration that need be made to the collision-free analysis is that ω must be replaced by $\omega - i\nu$.

For the present problem this approximation is not satisfactory, since it does not conserve particles locally. The particles tend to disappear where there is an excess and reappear where there is a deficiency. A more satisfactory approximation, as discussed by Dougherty,[32] is the BGK approximation[33] which does conserve particles locally and which is given by

$$(\partial f/\partial t)_c = -\nu[f_1(\mathbf{x}, \mathbf{v}, t) - [N_1(\mathbf{x}, t)/N_0]f_0(\mathbf{v})], \quad (14.3.8)$$

---

[31] B. D. Fried and S. D. Conte, "The Plasma Dispersion Function." Academic Press, New York, 1961.
[32] J. P. Dougherty, *J. Fluid Mech.* **16**, 126 (1963).
[33] P. L. Bhatnagar, E. P. Gross, and M. Krook, *Phys. Rev.* **94**, 511 (1954).

where N_1 and N_0 are the perturbation and background particle concentrations (the integrals of f_1 and f_0 over velocity space). With the collisions accounted for in this fashion, Dougherty found y to be given by

$$y = \frac{i + (\theta - i\psi)J}{1 - \psi J}, \qquad (14.3.9)$$

where J is the normalized Gordeyev integral [see (14.3.4)] including the effect of collisions; i.e.,

$$J = \int_0^\infty \exp[-i(\theta - i\psi)t' - \phi^{-2}\sin^2\alpha \sin^2 \tfrac{1}{2}\phi t' - \tfrac{1}{4}t'^2 \cos^2\alpha]\, dt' \qquad (14.3.10)$$

and ψ is the collision frequency (v) normalized as in (14.3.5)

$$\psi = (v/k)(m/2KT)^{1/2}. \qquad (14.3.11)$$

In the absence of a magnetic field, J can again be expressed in terms of the plasma dispersion function. Results similar to those of (14.3.9) and (14.3.10) have been obtained by Lewis and Keller.[34] Various approximate forms of (14.3.9) and numerical results for the power spectrum of the scattered signal are given by Dougherty and Farley.[27]

The effect of Coulomb (ion-ion) collisions on the shape of the spectrum has been examined in a crude semiquantitative fashion by Farley[35] and more rigorously by Dougherty[36] using an approximate Fokker–Planck equation. In the ionosphere ion-ion collisions are only important when **k** is very nearly perpendicular to the magnetic field **B**, in which case they often destroy the ion cyclotron resonance effect which is predicted by the collision-free theory. Dougherty's results are not very convenient to use in numerical calculations. The various integrals which must be evaluated numerically are related to the Gordeyev integral [(14.3.10) with ψ set equal to zero] and tend to diverge as $\alpha \to \pi/2$. Woodman,[37] starting from the same model Fokker–Planck equation, has avoided these difficulties and given detailed numerical results.

A current or a mean drift of the plasma as a whole can easily be accounted for in the theory by introducing a mean Doppler shift ($\theta \to \theta - \theta_0$, say) into the expressions for y_e or y_i or both. A drift of the plasma as a whole simply causes the spectrum to be Doppler shifted, without change of shape.

---

[34] R. L. Lewis and J. B. Keller, *Phys. Fluids* **5**, 1248 (1962).
[35] D. T. Farley, *J. Geophys. Res.* **69**, 197 (1964).
[36] J. P. Dougherty, *Phys. Fluids* **7**, 1788 (1964).
[37] R. F. Woodman, Tech. Rep. No. 3. Eng. Sci. Lab., Harvard Univ., Cambridge, Massachusetts, 1967.

14.3.2.2. Theoretical Results. In this section we shall briefly describe the way in which the various parameters of the plasma affect the total power and the power spectrum (or autocorrelation function) of the scattered signal.

14.3.2.2.1. TOTAL SCATTERING CROSS SECTION. The total scattered power is determined primarily by three parameters: the electron density (N), the ratio of Debye length to radar wavelength ($k^2\lambda_D^2$), and the temperature ratio T_e/T_i. If the temperature ratio is not unity, the power may also be affected somewhat by the magnetic field or collisions.

In some cases it is possible to obtain an analytical expression for the total scattering cross section, the integral of (14.3.2) over all Doppler shifts. We find that[29,38]

$$\sigma_{tot} = \int_{-\infty}^{\infty} \sigma(\omega_0 + \omega)\, d\omega$$

$$= N r_e^2 \sin^2 \delta \left[\frac{k^2 \lambda_D^2}{1 + k^2 \lambda_D^2} + \frac{1}{(1 + k^2 \lambda_D^2)(1 + T_e/T_i + k^2 \lambda_D^2)} \right].$$

(14.3.12)

The two terms inside the square brackets correspond to the so-called electronic and ionic components of the spectrum.

When the electron and ion temperatures are equal, this result is exact, and the quantity inside the brackets becomes $(1 + k^2 \lambda_D^2)/(2 + k^2 \lambda_D^2)$ and reduces to $\frac{1}{2}$ when the Debye length is small; the total scattered power is just half that predicted by the simple Thomson scatter theory. When the Debye length is large, this factor approaches unity and we recover the Thomson result for scattering from completely free electrons with independent trajectories. These results are unaffected by collisions or the presence of a magnetic field.

When the electron and ion temperatures are not equal, (14.3.12) is only approximately true.[21,29,39] The cross section does not continue to decrease as the ratio T_e/T_i increases; at some point it begins to increase, and for very large ratios the term in square brackets in (14.3.12) becomes unity. The ratio at which the approximation begins to break down depends on the electron-to-ion mass ratio, the magnetic field, and sometimes the collision frequency (although T_e/T_i is unlikely to differ much from unity when collisions are important). This point is illustrated in Fig. 7 for the common case in which the Debye length, the electron Larmor radius, and the collision frequency are negligibly small. It can be seen that

---

[38] O. Buneman, *J. Geophys. Res.* **67**, 2050 (1962).
[39] D. R. Moorcroft, *J. Geophys. Res.* **68**, 4870 (1963).

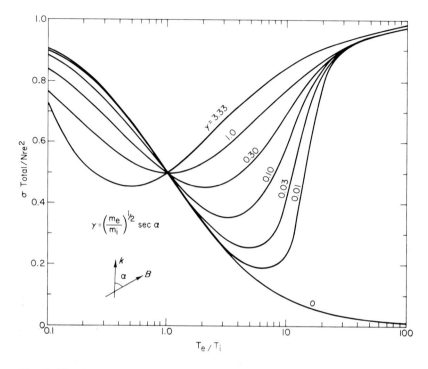

FIG. 7. The dependence of the scattering cross section for incoherent scatter on the electron-to-ion temperature ratio and $\gamma = (m_e/m_i)^{1/2} \sec \alpha$, where α is the angle between **k** and the magnetic field **B**, and $(m_e/m_i)^{1/2} \approx 6 \times 10^{-3}$ in the F region. The Debye length is assumed to be negligibly small. The curve for $\gamma = 0$ is just $(1 + T_e/T_i)^{-1}$. [From D. T. Farley, *J. Geophys. Res.* **71**, 4091 (1966).]

the approximation $\sigma_{tot} \propto (1 + T_e/T_i)^{-1}$ is good only for rather small values of the temperature ratio. Nevertheless, since this ratio seldom exceeds values of 3 or 4 in the ionosphere, the approximation is often adequate. It cannot be used for observations near the magnetic equator, however, for which the angle α is close to 90°.

14.3.2.2.2. POWER SPECTRUM. The shape of the power spectrum or autocorrelation function of the scattered signal depends in a complicated way on the various plasma parameters. As a result, the analysis of the data is sometimes difficult; on the other hand, when the analysis is properly done it yields a great deal of information about the scattering medium. The important factors are Debye length, electron and ion temperature, major ionic constituents, collision frequencies, magnetic field strength and orientation, plasma drift velocity parallel to **k**, current strength parallel to **k**, and high energy "tail" of electron velocity distribution.

14.3.2.2.2.1. *Large Debye Length.* The simplest but perhaps least interesting case to consider is that of large Debye length ($k^2 \lambda_D^2 \gg 1$). The electrons then act as free independent particles and the spectrum is Gaussian shaped with Doppler shifts characteristic of the mean thermal velocity of the electrons. If **k** is nearly perpendicular to **B** and the electron Larmor radius is sufficiently large ($\phi_e \lesssim 1$), the spectrum will be modulated at the electron cyclotron frequency; i.e., it will consist of a series of peaks separated by the cyclotron frequency and having a Gaussian envelope. This case of large Debye length is of little practical interest in ionospheric work because of the wide bandwidth of the returned signal and the resulting low signal-to-noise ratio.

14.3.2.2.2.2. *The Plasma Line.* When the Debye length is small ($k^2 \lambda_D^2 \ll 1$) the spectrum splits into two components, generally called the electronic and ionic. The electronic component consists usually of two spectral lines, one above and one below the operating frequency, and separated from it by approximately the local plasma frequency, ω_p. The actual separation, ω_r, depends slightly on thermal and magnetic effects. If $\Omega_e^2 \ll \omega_p^2$, it can be approximated by[24,40]

$$\omega_r^2 = \omega_p^2(1 + 3k^2\lambda_D^2) + \Omega_e^2 \sin^2 \alpha. \qquad (14.3.13)$$

If the velocity distribution of the electrons is Maxwellian, with a typical ionospheric temperature of perhaps a few thousand degrees or less, the plasma lines contain very little power (a scattering cross section of $\sigma_{tot} k^2 \lambda_D^2$) and are too weak to be detectable with radars now in use. However, during the day in the F region of the ionosphere, the plasma contains a high energy "tail" of energetic (up to perhaps 25–30 eV) photo-electrons which are created by the solar ionizing radiation and which only slowly become thermalized. Although these electrons make up only a small percentage of the total, they can cause an enhancement in the plasma line return by a factor of as much as 50. When magnetic effects are not important, the enhancement factor I is given by[40]

$$I = \frac{f_{th}(v_\phi) + f_p(v_\phi) + \chi}{f_{th}(v_\phi) - (KT/mv_\phi)f_p'(v_\phi) + \chi}, \qquad (14.3.14)$$

where f_{th} and f_p are the one-dimensional (in the direction of **k**) velocity distributions of the thermalized electrons and the photoelectrons, v_ϕ is the phase velocity ω_r/k of the plasma wave, T is the temperature of the thermal component, and the prime denotes a derivative with respect to

---

[40] F. W. Perkins and E. E. Salpeter, *Phys. Rev. A* **139**, 55 (1965).

velocity. The factor χ describes the contribution of electron-ion collisions to the damping and excitation of plasma waves and is given by

$$\chi = (k^3/6N\pi^2)(m/2\pi KT)^{1/2} \ln(4\pi N\lambda_D^3). \qquad (14.3.15)$$

When v_ϕ is small the f_{th} terms in (14.3.14) dominate and I is approximately unity (no enhancement); when v_ϕ corresponds to energies much larger than KT, the f_p terms may dominate and I becomes approximately T_p/T, where T_p is the "temperature" of the photoelectrons, as defined by the slope of the velocity distribution at $v = v_\phi$. For very large phase velocities neither the thermal nor the photoelectrons will interact coherently with the wave. Only the collision term will remain and I will again reduce to unity.

The enhancement thus takes place only in a certain range of values of v_ϕ, which in turn corresponds to a range of plasma frequencies determined by the radar wavelength. In the ionosphere a useful rule of thumb is that significant enhancements for backscatter will occur for plasma frequencies of roughly $2/\lambda$ to $6/\lambda$, where λ is the radar wavelength in meters, and the plasma frequency is measured in megahertz. Radar operating frequencies of 300–500 MHz are thus optimum for most measurements of these effects.

By measuring the frequency of the plasma line resonance as a function of altitude, an accurate, unambiguous determination of the electron density as a function of altitude may be obtained. By measuring the strength of the plasma line as a function of frequency it is possible to deduce quite a bit about the photoelectron energy distribution. Plasma line observations are possible not only during the day, but also during at least part of the night during local winter when the magnetically conjugate point is sunlit. A significant number of photoelectrons are able to travel along the magnetic field lines and traverse the entire magnetosphere without serious energy loss.[41-44]

The Landau damping terms, and hence the enhancement factor (14.3.14), can be affected by the presence of a magnetic field.[42] The magnetic field becomes increasingly important as **k** approaches perpendicularity with **B**, and the enhancement can be reduced drastically when the resonant frequency is an integral multiple of the cyclotron frequency. Successful observations of these effects[42,45] have provided an excellent verification

---

[41] H. C. Carlson, *J. Geophys. Res.* **71**, 195 (1966); *Radio Sci.* **3**, 668 (1968).
[42] K. O. Yngvesson and F. W. Perkins, *J. Geophys. Res.* **73**, 97 (1968).
[43] H. Carru, M. Petit, and P. Waldteufel, *J. Atmos. Terr. Phys.* **29**, 351 (1967).
[44] J. V. Evans, *J. Geophys. Res.* **73**, 3489 (1968).
[45] E. J. Fremouw, J. Petriceks, and F. W. Perkins, *Phys. Fluids* **12**, 869 (1969).

of Landau damping theory. Perkins[46] has also predicted that an additional spectral line at about the lower hybrid frequency, $(\Omega_e \Omega_i)^{1/2}$, might be observable under certain conditions and could be used to determine the mean ionic mass. The necessary criteria for its observation are not very well satisfied at observatories now in operation, however, and so far it has not been seen.

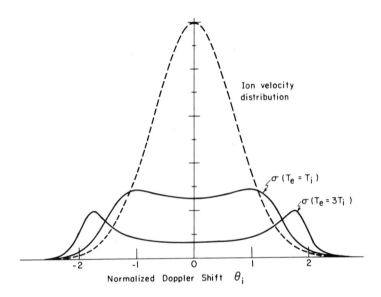

FIG. 8. The frequency dependence of the scattering cross section. The frequencies are normalized by the ion thermal velocity; i.e., $\theta_i = (\omega/k)(m_i/2KT_i)^{1/2}$. The dashed curve corresponds to hypothetical scattering by free ions with the free electron scattering cross section.

14.3.2.2.2.3. *The Ionic Component.* The second and probably most important component of the spectrum (in the small Debye length case) is the ionic component. Here the Doppler shifts are characteristic of the ion thermal velocities, and the signal bandwidths are of the order of tens of kilohertz or less for typical ionospheric conditions and radar operating frequencies. Two examples of the shape of the ion component are given in Fig. 8. We assume that only a single ion species is involved and that the effects of the Debye length, magnetic field, collisions, etc., are negligible.

The dashed curve is for comparison only and shows the spectrum that would be obtained in a hypothetical situation in which the ions did the

---

[46] F. W. Perkins, MATT-555. Plasma Phys. Lab., Princeton Univ. Princeton, New Jersey, 1967.

scattering and behaved as completely free particles with the electron Thomson scattering cross section. The area under the dashed curve is just twice the area (total scattering cross section) under the solid curve for $T_e = T_i$. The spectrum has a characteristic double-humped shape with a depression in the center. When $T_e = T_i$ the depression is slight and the peaks occur at Doppler shifts of $\theta_i = \pm 1$, or

$$\omega = \pm k(2KT/m_i)^{1/2}. \qquad (14.3.16)$$

The half-power points are at approximately $\theta_i = \pm 1.6$. In this normalized form the spectral shape is nearly independent of the ion mass. The dependence becomes important only when terms of order $(m_e/m_i)^{1/2}$ are comparable to unity. (The mass may be important when the magnetic field is introduced, however.) Of course the actual signal bandwidth is proportional to $m_i^{-1/2}$. The shape of the autocorrelation function for $T_e = T_i$ is shown in Fig. 9. It is a purely real function when the spectrum is symmetric about the operating frequency.

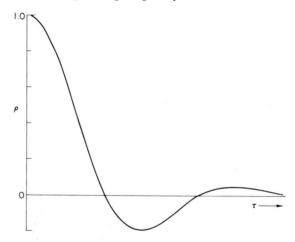

FIG. 9. A typical autocorrelation function $\rho(\tau)$ for incoherent scatter from a collisionless plasma in thermal equilibrium containing a single ion species and no magnetic field.

When the electron temperature is greater than the ion temperature, as it often is in the ionosphere, the spectrum becomes more sharply double peaked, as is also shown in Fig. 8, while the total scattered power decreases, as mentioned previously. Moorcroft[47] and Farley[29] have given further numerical results and pointed out that the position of the peak in the spectrum and of the zero-crossing in the autocorrelation function

---

[47] D. R. Moorcroft, *J. Geophys. Res.* **69**, 955 (1964).

14.3. INCOHERENT SCATTERING

are largely a function of T_e, and depend only to a slight extent on T_i. The ratio T_e/T_i can be determined easily from the "peak-to-valley" ratio in the spectrum or the depth of the first negative loop in the autocorrelation function.

Moorcroft[47] has shown that the effect of a finite Debye length on these results can be accounted for to a good approximation in a very simple way. If the electron temperature is replaced by a fictitious temperature

$$T_e' = T_e(1 + k^2\lambda_D^2)^{-1}, \qquad (14.3.17)$$

the change in the shape of the theoretical curves calculated for $k^2\lambda_D^2 = 0$ is barely perceptible, even for values of $k^2\lambda_D^2$ of order unity. The total scattered power *is* changed, however [see (14.3.12)].

The shape of the spectrum is of course affected by the presence of ion mixtures. The only important ion at 300 km in the ionosphere is O^+, but below about 200 km heavier ions (mainly NO^+ and O_2^+) gradually begin to predominate, and at altitudes well above 300 km H^+ and perhaps He^+ become important. Moorcroft[47] and others have made numerical calculations of the spectral shapes. In practice it is usually difficult to separate the effect of a small change in composition from that of a small change in T_e/T_i. This is particularly true for O^+-heavy ion and O^+-He^+ mixtures. In order to obtain the composition from the spectral data, the temperature ratio must be known or assumed, and vice versa. Generally speaking, no more than two or three independent parameters of the plasma can be determined accurately from the ionic component of the spectrum.

Depending upon the latitude, altitude, and time of day, it may be possible to make the necessary assumptions with a good deal of confidence; in other cases it may not. At the magnetic equator, for example, the temperature ratio differs from unity only during the day at altitudes where the only important ion is atomic oxygen. At other latitudes, however, the problem is more difficult. It is sometimes possible to use plasma line and total scattered power measurements to determine the temperature ratio independently [using (14.3.12)] or to use rocket or satellite measurements of the composition. An added complication is introduced when the Debye length is not small.

Collisions between ions and neutral particles become important when the ionic mean free path becomes comparable to k^{-1} (i.e., when $\psi_i \sim 1$). For radar frequencies of the order of 400 MHz, this begins to happen at altitudes below about 120 km. Dougherty and Farley[27] have shown how the spectrum changes from double-humped to bell shaped as the normalized collision frequency is increased. When the plasma is collision dominated ($\psi_i \gg 1$), only a single parameter, the diffusion coefficient, can be determined from the spectrum. Tanenbaum[20] has pointed out that under certain conditions an additional resonance line, corresponding to sound waves in the *neutral* gas, may be detectable in the collision-dominated case. The necessary conditions are unlikely to be met in the ionosphere, however.

The magnetic field can have an important effect on the ionic component if **k** is very nearly normal to **B**. The spectrum becomes narrower and the double hump gradually disappears, a prediction which has been verified by Baron and Petriceks.[48] If k^{-1} is smaller than the ion Larmor radius ($\phi_i < 1$) an ion cyclotron resonance effect may be observable.[28] This resonance would seem capable of providing a convenient way of determining the ionic composition. The resonance could be observed for each ion present and the relative strengths would give the composition, just as in a mass spectrometer. In fact, it has now been realized[35-37] that ion-ion collisions will normally destroy the resonance in the ionosphere for all ions except protons. The effect has been observed for protons, however, and a

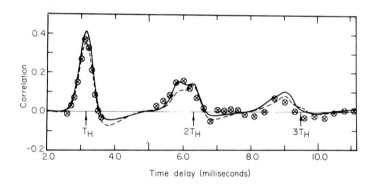

FIG. 10. A comparison of theory with a measurement of the proton gyroresonance. The correlation function is shown only at long delays. Theoretical curves for two assumed compositions are given. The agreement could be improved further by a slight change in the assumed value of magnetic field strength. [From D. T. Farley, *Phys. Fluids* **10**, 1584 (1967).] Key: (—) 30% O^+, 70% H^+; (– – –) 50% O^+, 50% H^+; × data, 700 km.

comparison of theory and experiment is shown in Fig. 10.[49] The agreement is excellent and could be improved still further by a slight change in the assumed value of the magnetic field.

If the plasma as a whole is drifting with a velocity component parallel to the direction of the radar beam (i.e., parallel to **k**) the entire spectrum will undergo a corresponding Doppler shift. Since typical velocities are of

---

[48] M. Baron and J. Petriceks, *J. Geophys. Res.* **72**, 5325 (1967).
[49] D. T. Farley, *Phys. Fluids* **10**, 1584 (1967).

the order of a few tens of meters per second, the Doppler shifts are quite small and may be only one percent or so of the width of the spectrum. The effects are difficult to detect, therefore, but nevertheless drifts are now being measured routinely at several observatories to accuracies of the order of ±1–10 m/sec. The measurement becomes easier when the spectrum is narrowed by the effect of collisions or a magnetic field, so that the mean Doppler shift becomes a larger fraction of the Doppler spread. Woodman and Hagfors[50] have taken advantage of the magnetic effect to make particularly accurate drift measurements at the magnetic equator (Jicamarca).

In the presence of a mean drift $\mathbf{V}_d$ the correlation function becomes complex, i.e.,

$$\rho(\tau) \to \rho(\tau) \exp[-i\mathbf{k} \cdot \mathbf{V}_d \tau]. \qquad (14.3.18)$$

The drift velocity can be derived by measuring the phase shift term.

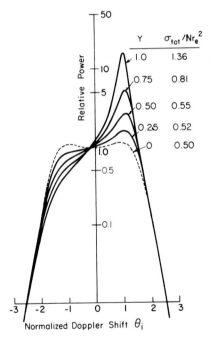

FIG. 11. The dependence of the incoherent scatter spectrum on the strength of the current parallel to $\mathbf{k}$. Thermal equilibrium is assumed and $k^2 \lambda_D^2$ is neglected. $Y = v_e(m_e/KT)^{1/2}$ is the relative drift velocity of the electrons normalized by their thermal velocity. [From M. N. Rosenbluth and N. Rostoker, *Phys. Fluids* **5**, 776 (1962).]

[50] R. F. Woodman and T. Hagfors, *J. Geophys. Res.* **74**, 1205 (1969).

Currents in the direction of **k** cause the spectrum to become asymmetrical; one of the peaks or shoulders (the one corresponding to the direction in which the electrons are flowing) is enhanced, the other depressed. The total scattered power is also increased. These effects are illustrated in Fig. 11, which is adapted from the work of Rosenbluth and Rostoker.[30] Thermal equilibrium has been assumed and the effects of magnetic field, Debye length, collisions, etc. have been neglected. A corresponding change occurs in the correlation function; it becomes complex, but the phase shift will not be linear in τ as it is in (14.3.18). The effect of a current is more pronounced if the electron-to-ion temperature ratio is greater than unity. If the current becomes large enough, of course, the plasma becomes unstable (two-stream instability). Since T_e/T_i never gets much larger than 3 or 4 in the ionosphere, however, appreciable current effects can occur only when the relative drift velocity is a significant fraction of the electron thermal velocity. Such strong currents do not exist in the ionosphere and this effect has so far gone unobserved.

14.3.2.2.3. SUMMARY. Let us summarize the main theoretical results for typical ionospheric conditions. Besides the obvious dependence on the electron density, the total scattered power or total scattering cross section is most seriously affected by the electron-to-ion temperature ratio. The dependence on magnetic field (for $T_e \neq T_i$) and Debye length must be kept in mind, but these parameters are generally less important in practice. The dependence on collision frequency (again for $T_e \neq T_i$) seems to be of only academic interest at the moment.

Many more parameters of the plasma affect the shape of the spectrum and autocorrelation function, and it is not always easy to separate the various effects. Fortunately it is usually the case that only a few are significant in a particular region of space. The temperature ratio is usually close to unity when collisions are important, for example, and the magnetic field need be considered only if **k** is very nearly orthogonal to **B**. The temperature ratio and the composition are certainly the most important parameters in much of the ionosphere; the Debye length becomes significant for high radar frequencies and low electron densities; and collisions become dominant at low altitudes. The plasma line component of the spectrum depends primarily on the local plasma frequency and the photoelectron energy spectrum, but it can also be influenced by the magnetic field.

14.3.3. Experimental Observations

It is beyond the scope of this part to go into a detailed discussion of all that has been learned about the ionosphere from incoherent scattering

14.3. INCOHERENT SCATTERING

measurements. Some results have already been mentioned; a complete treatment and full list of references can be found in the review articles already referred to.

From the point of view of the plasma physicist, the most significant feature of the ionospheric measurements is their remarkable agreement with theory. It is probably not overstating the case to say that they have provided the best quantitative test to date of the usual linearized kinetic theory. No observation has been made which is in any way inconsistent with the present theory, and all but a very few of the predictions of the theory have been experimentally verified, sometimes to accuracies of the order of one percent. The effect of the temperature ratio, the Debye length, and the magnetic field on both the total scattered power and the spectrum have been verified. The effects of composition and collisions, both two particle (ion-neutral) and small angle (ion-ion), have been clearly observed as has the enhancement of the plasma line by nonthermal electrons and the dependence of the plasma line on the magnetic field.

In some cases it has been possible to make internal checks on the consistency of the data. For example, Eq. (14.3.12) can be used to compute the temperature ratio T_e/T_i from power measurements if the electron density is known and the Debye length is small. The electron density can be determined simultaneously and independently from plasma line or Faraday rotation data. The resulting values of the temperature ratio can be compared with values deduced from simultaneous measurements of the power spectrum or correlation function. Such comparison have given excellent agreement.

Independent measurements of electron density, temperature, composition, etc. can be made using *in situ* rocket and satellite probes or other ground based techniques, such as conventional ionosondes. The results have usually agreed well with simultaneous incoherent scatter observations, with the possible exception of electron temperature measurements. A number of satellite measurements in the F region have given values of T_e consistently higher (sometimes by as much as 70%) than those obtained from incoherent scatter.[51] At the time of writing, this discrepancy is still unresolved, but the weight of evidence seems to favor the incoherent scatter values.[17]

Practically all of the important parameters of the ionospheric plasma can be measured using incoherent scatter. The resulting comprehensive local picture of the ionosphere complements the cruder but global coverage provided by satellites. Our understanding of the production and loss of

---

[51] W. B. Hanson, L. H. Brace, P. L. Dyson, and J. P. McClure, *J. Geophys. Res.* **74**, 400 (1969).

ionization in the ionosphere, the temperature balance, the coupling between hemispheres and the protonosphere along magnetic field lines, and dynamic effects due to such things as gravity waves, to cite a few examples, has advanced considerably in recent years because of incoherent scatter measurements. The temperature of the neutral atmosphere at E and F region altitudes can be accurately determined from measurements of T_i, T_e, and the electron density.

The highest altitude at which power measurements are presently feasible is about 10,000 km, while the lowest is about 80 km. Spectrum or correlation function measurements are presently made to about 1500 km, but probably could be pushed a bit higher. The best altitude resolution usually obtained in the E or F region is about 3–5 km, but the use of pulse compression techniques should make it possible to reduce this figure to 1 km or less.

14.4. Scattering from the Equatorial Electrojet

14.4.1. Introduction

In the preceding section we were concerned with scattering from very weak thermal fluctuations in the electron density. Now we turn our attention to much stronger effects in which the scattering cross section is larger by factors of as much as 10^8. We shall restrict ourselves to effects caused by currents in the E region of the ionosphere, although scattering of comparable strength can be obtained at times from the F region. The F region scatter, usually known as spread F, has been reviewed by Herman.[52] The physical mechanisms responsible for it are not at all well understood at present.

A world-wide system of electric fields and currents exists in the ionosphere. The driving force is the dynamo action of the atmospheric tides in conjunction with the geomagnetic field. The motion of the neutral atmosphere drags the charged particles across the lines of force, and the resulting charge separation produces the electric fields. The horizontal current flow is confined to the E region, where the conductivity is greatest. Current can of course also flow along the lines of force, but such currents are quite weak (except perhaps at auroral latitudes) and are not of concern here. The E region currents are particularly strong at the geomagnetic equator and in an auroral arc, and strong radar returns are obtained from both regions. The characteristics of the echoes from the two regions are quite similar in many respects, but the auroral data are more complex and variable, and

[52] J. R. Herman, *Rev. Geophys.* **4**, 255 (1966).

hence are more difficult to unravel. It is paradoxical but probably true that the best way to begin to understand the auroral phenomena is to study the effects at the equator.

As we shall see, at least some of the equatorial observations can be explained by a two-stream instability theory; when the current reaches a certain threshold level the plasma becomes unstable and strong acoustic waves are generated with phase fronts aligned along the magnetic field. Those with the proper orientation and wavelength (proper **k**) can give rise to very strong radar echoes.

14.4.2. The Electrojet

The motion of the ions and electrons in the presence of constant electric and magnetic fields can be described in terms of mobility and conductivity tensors. If we follow Rishbeth and Garriott[53] and define the mobility to be the ratio of a velocity to an applied force, i.e., a mobility per unit charge, we have

$$k_0 = \frac{1}{m\nu}, \quad k_1 = \frac{\nu^2}{\nu^2 + \Omega^2} k_0, \quad k_2 = \frac{\nu\Omega}{\nu^2 + \Omega^2} k_0, \quad (14.4.1)$$

where k_0 is the direct or longitudinal mobility ($\|\mathbf{B}, \|\mathbf{E}$), k_1 is the transverse or Pedersen mobility ($\perp\mathbf{B}, \|\mathbf{E}$), k_2 is the Hall mobility ($\perp\mathbf{B}, \perp\mathbf{E}$), and ν and Ω are the appropriate collision and cyclotron frequencies for the ions or electrons. The collisions are primarily with neutral particles. In terms of these mobilities, the conductivities in the ionosphere are given by

$$\sigma_0 = Ne^2(k_{0e} + k_{0i}), \quad \sigma_1 = Ne^2(k_{1e} + k_{1i}),$$
$$\sigma_2 = Ne^2(k_{2e} - k_{2i}), \quad \sigma_3 = \sigma_1 + \sigma_2^2/\sigma_1. \quad (14.4.2)$$

(Hopefully the use of σ for both conductivity and scattering cross section will cause no confusion.) Since k_{2e} is always greater than k_{2i}, the Hall current always flows in the direction $\mathbf{B} \times \mathbf{E}$. The last expression in (14.4.2) defines the Cowling conductivity, which turns out to be the effective conductivity in the electrojet region and is only slightly smaller than σ_0.

At the height of the electrojet, the Hall conductivity exceeds the Pederson conductivity by a factor of ten or more. The predominantly east–west electric field thus tries to create a vertical current at the equator. This current is prevented from flowing by the insulating layers above and below, where the Hall conductivity is small. A vertical polarization field

---

[53] H. Rishbeth and O. K. Garriott, "Introduction to Ionospheric Physics." Academic Press, New York, 1969.

is thus set up which is stronger by a factor of σ_2/σ_1 than the original driving field. This induced vertical polarization field now drives a horizontal Hall current. The overall result is a ratio of horizontal current to horizontal electric field given by σ_3.

The electrojet current is a maximum at an altitude of about 105 km, where $v_e \ll \Omega_e$, $v_i \gg \Omega_i$, and $v_e v_i \sim \Omega_e \Omega_i$. The ions are essentially at rest with respect to the neutrals. The main current-carrying region extends over about 10 km in altitude and a few hundred kilometers in latitude. During the day the orders of magnitude of some of the important electrojet parameters are (in the units commonly used):

electron density	$\sim 10^5/\text{cm}^3$
magnetic field	~ 0.3 gauss
E-W electric field	$\sim 10^{-3}$ volt/meter
electron drift velocity	~ 500 meter/sec
ion thermal velocity	~ 350 meter/sec
current density	$\sim 10^{-5}$ ampere/meter^2
height integrated current density	$\sim 10^2$ amperes/km

During the night the electron density and hence the current strength are smaller by perhaps two orders of magnitude. The electric field strength and electron drift velocity may not be reduced, however. The conventional current is toward the east during the day, and hence the electrons drift towards the west. The directions are reversed at night. Recent numerical models of the electrojet have been given by several authors.[54-56]

14.4.3. VHF Scatter Observations

Radar echoes have been observed from the aurora ever since the early days of World War II,[3,57] but the equatorial measurements that we are primarily concerned with are more recent. It was recognized some time ago that a certain type of scatter return on a conventional ionogram (essentially the output of a swept frequency hf radar) was closely associated with changes in the equatorial magnetic field, and hence the electrojet

---

[54] M. Sugiura and J. C. Cain, *J Geophys. Res.* **71**, 1869 (1966).
[55] J. Untiedt, *J. Geophys. Res.* **72**, 5799 (1967).
[56] M. Sugiura and D. J. Poros, *J. Geophys. Res.* **74**, 4025 (1969).
[57] H. G. Booker, *in* "Physics of the Upper Atmosphere" (J. A. Ratcliffe, ed.), Chapter 8. Academic Press, New York, 1960.

current.[58] This hf scatter is generally called equatorial sporadic E. However, the real progress which has been made recently in our understanding of the physics of the scattering phenomena has been due primarily to the results of the vhf scattering measurements carried out in Peru at Jicamarca.[59-65]

Most of the measurements were made at a frequency of 50 MHz, but some data at 148 MHz are also available. Relatively low peak transmitter powers (~ 5 kW) were usually used at 50 MHz. The antenna was directed normal to the magnetic field and was steerable in the east–west direction only. The power and the autocorrelation function of the scattered signal were measured. As many as 64 delays were used in the latter measurement, and the result was Fourier transformed to obtain the power spectrum. The observations are summarized below.

14.4.3.1. Total Power. The strength of the total scattered signal depends on several quantities, the most important of which are (1) the strength of the electrojet current or electron drift velocity, (2) the orientation of the radar beam with respect to the magnetic field, and (3) the orientation with respect to the zenith. Strong echoes do not appear during the day unless the current strength exceeds some minimum threshold value. The threshold appears to be actually in the electron velocity rather than the current density, because strong echoes can also be present at night when the electron density, and hence the current strength (but not the electron velocity), is quite small. The recent work of Balsley[65] suggests that there actually may be two thresholds. Relatively weak echoes appear when the first is exceeded, while much stronger signals are observed after the second is crossed.

The echoes are aspect sensitive; i.e., they only appear if the radar is directed to within a degree or so of orthogonality to the earth's magnetic field. The evidence indicates that the echoes are from traveling plane wave irregularities whose phase fronts are very closely aligned with the magnetic field.

[58] S. Matsushita, *J. Geomagn. Geoelec.* **3,** 44 (1951); *in* "Ionospheric Sporadic E" (E. K. Smith, Jr., and S. Matsushita, eds.), p. 334. Pergamon, New York, 1962.

[59] K. L. Bowles, R. Cohen, G. R. Ochs, and B. B. Balsley, *J. Geophys. Res.* **65,** 1853 (1960).

[60] K. L. Bowles, B. B. Balsley, and R. Cohen, *J. Geophys. Res.* **68,** 2485 (1963).

[61] K. L. Bowles and R. Cohen, *in* "Ionospheric Sporadic E" (E. K. Smith and S. Matsushita, eds.), p. 51. Pergamon Press, New York, 1962.

[62] R. Cohen and K. Bowles, *J. Geophys. Res.* **68,** 2503 (1963); *J. Res. Nat. Bur. Stand. Sect. D* **67,** 459 (1963).

[63] R. Cohen and K. L. Bowles, *J. Geophys. Res.* **72,** 885 (1967).

[64] B. B. Balsley, *J. Geophys. Res.* **70,** 3175 (1965).

[65] B. B. Balsley, *J. Geophys. Res.* **74,** 2333 (1969).

The strongest scattering (after removing geometrical factors) is at low elevation angles. In other words the strongest irregularities (plasma waves) are those traveling in directions most nearly parallel to the direction of the electron flow. The scattering cross section is sometimes of the order of 10^8 times as strong as the cross section for incoherent scatter. Echoes obtained from overhead are much weaker than oblique echoes, but they are not zero. Measurements with a vertically directed radar using short pulses have shown that the scattering region usually extends over about 10 km in altitude, centred near 105 km. This is also the region in which the current is known to be strongest. When the electrojet is building up or dying away and the scatter is weak, the scattering region is often observed to bifurcate; the echoes come from the edges of the current-carrying region where the current gradients are greatest. Even when the electrojet is strong the scattering is usually "patchy"; in other words, even though on the average the scatter is from the entire altitude range of the electrojet, at any given instant it comes primarily from a few large patches which have lifetimes of a few tens of seconds, dimensions of the order of a kilometer, and an appreciable drift velocity.

Such observations as exist at 148 MHz are consistant with the 50 MHz data. The power measurements by themselves already suggest that some sort of instability mechanism may be important. This conclusion is strengthened by the spectral observations.

14.4.3.2. Frequency Spectrum. A number of characteristics of the power spectrum of the scattered signals are illustrated in Figs. 12–15, which are taken from the work of Cohen and Bowles.[63] All the spectra shown are normalized to the same maximum value, and so the areas under the curves do *not* indicate the relative strengths of the signals. In Fig. 12, for example, the signal for the vertically directed antenna is much weaker than for the other cases, even though the area under the spectrum shown is larger.

Figure 12 shows a series of observations looking in different directions for a time when the electrojet and the scattering were strong. It can be seen that the results pointing east and west are nearly mirror images of each other, and that the position of the Doppler peak in the spectrum does not change when the zenith angle is shifted from 70° to 45°. Recalling that radar echoes can only be obtained from waves with the proper wave vector **k** [see Eq. (14.2.3)], we deduce that the oblique echoes all correspond to waves with a wavelength of 3 meters traveling at least partially in the westward direction, the direction of electron flow in the daytime. The phase velocity of the waves does not depend on the direction of propagation, and for Fig. 12 is about 400 meters/sec, which is slightly greater than the acoustic or ion thermal velocity $(2KT/m_i)^{1/2}$ in the electrojet. The spectrum

14.4. SCATTERING FROM THE EQUATORIAL ELECTROJET

of the weak echo from overhead, which corresponds to vertically traveling waves, is much more variable. These waves are not understood at present, and we will devote most of our attention to the stronger, oblique scatter. As we shall see shortly, many features of the obliquely traveling waves can be explained on the basis of a two-stream plasma instability theory.

In Fig. 13 we have an example of how the spectrum can change with time (and electrojet strength) for a fixed direction of observation. When the current (as indicated by the geomagnetic field variations) is fairly weak, the spectrum does not resemble those of Fig. 12. The weak scattered signal has a more diffuse spectrum with a much smaller mean Dopper shift. As the current strength (electron velocity) increases, so does the strength of the scattered signal (although this is not shown on the figure because of the normalization procedure), and the mean Doppler shift moves to higher values. Balsley[65] has shown that the mean Doppler shift of these weak echoes can be used to deduce the horizontal electron drift velocity, and that the square of this velocity is proportional to the scattered power. With further velocity increases, the peak at the acoustic velocity appears and finally dominates the spectrum.

The weak irregularities are referred to in the literature as "secondary," "type II," or "non–two-stream" irregularities. If the normalization factors are properly taken into account, one finds that the type II irregularities do not become weaker as the electron drift velocity increases, as Fig. 13 seems to suggest. They just become insignificant compared to the much stronger type I, or two-stream irregularities.

Figure 14 shows how the shape of the spectrum, and the relative proportion of type I and II scatter, can vary as a function of zenith angle β. The type I echoes become more pronounced as the radar beam is directed more nearly parallel to the direction of electron flow.

In Fig. 15 we have an illustration of the east–west asymmetry that sometimes is present in the measurements. The westward looking measurement is sandwiched between the two observations to the east to reduce any effects of local time variations. It can be seen that the position of the peak in the spectrum is symmetric, but the relative strengths of the type I and type II echoes are not. The total scattered power is usually somewhat greater when looking towards the west (upgoing plasma waves). Some indication of this same effect can also be observed in the central curve of Fig. 12. This asymmetry has no explanation at present.

14.4.4. Theory

The threshold effect observed in the power measurements, particularly of the oblique echoes, suggests strongly that the scatter is associated with some sort of instability process. The spectrum measurements

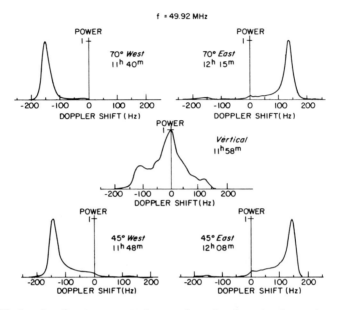

FIG. 12. A series of power spectra of scatter from the electrojet observed near noon at Jicamarca, Peru at various angles east and west of vertical. These spectra correspond to a relatively strong electrojet. [From R. Cohen and K. L. Bowles, *J. Geophys. Res.* **72**, 885 (1967).]

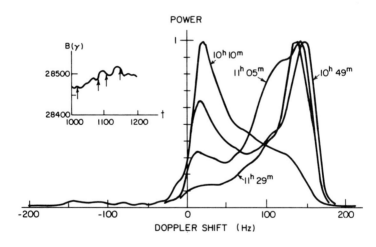

FIG. 13. The variation of the scatter spectrum as the strength of the electrojet changes. The observations were all at a zenith angle of 50° E. The variations in the electrojet current strength are indicated by the magnetic field measurements ($1\gamma = 10^{-5}$ gauss); the arrows indicate the times at which the power spectra were obtained. [From R. Cohen and K. L. Bowles, *J. Geophys. Res.* **72**, 885 (1967).]

14.4. SCATTERING FROM THE EQUATORIAL ELECTROJET

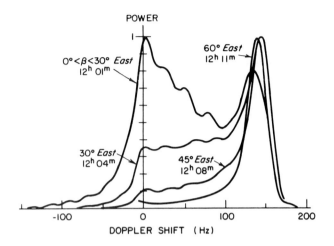

FIG. 14. Power spectra for a relatively strong and steady electrojet for various zenith angles. [From R. Cohen and K. L. Bowles, *J. Geophys. Res.* **72,** 885 (1967).]

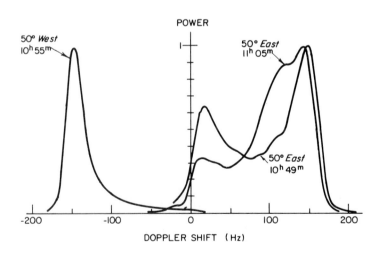

FIG. 15. An illustration of the east–west asymmetry sometimes observed in the scatter from the electrojet. The zenith angles were the same in each direction, and the electrojet was steady but not too strong. [From R. Cohen and K. L. Bowles, *J. Geophys. Res.* **72,** 885 (1967).]

indicate that acoustic waves are being generated in the plasma by the streaming electrons. Given these clues, the simple two-stream plasma instability comes immediately to mind. Unfortunately, however, the simple theory requires streaming velocities of the order of the *electron* thermal velocity before the medium becomes unstable (unless $T_e \gg T_i$, which is not the case here), whereas the actual velocities are only somewhat greater than the *ion* thermal velocity. The ordinary two-stream theory does not include magnetic effects, however, whereas in the electrojet we have a rather unusual situation in which a large current is flowing normal to the magnetic field.

Farley[66] developed a two-stream instability theory appropriate to the electrojet environment and showed that it did in fact account quantitatively for many features of the observations. The derivation was based on the Vlasov equation and took account of collisions by means of a BGK expression [see Eq. (14.3.8)] which conserves particles locally. The results show that waves traveling normal to the magnetic field will become unstable when $\mathbf{k} \cdot \mathbf{V}_d / |k|$, the component of drift velocity in the direction of the wave, exceeds the ion thermal velocity by a small amount. If the orthogonality condition for $\mathbf{k}$ and $\mathbf{B}$ is violated by more than a degree or so, the threshold velocity is increased substantially.

These instability and orthogonality conditions agree well with the observed threshold velocity and aspect sensitivity of the echoes. Moreover, the phase velocity of the marginally unstable waves predicted by theory matches the observed peak in the power spectrum; i.e., the value of Re(ω) predicted when Im(ω) → 0 is the value observed at the maximum. Since it is the component of the drift in the direction of $\mathbf{k}$ that excites the wave, we expect the waves traveling most nearly parallel to the electron mean drift velocity (and of course in the direction of the drift) to be the most strongly excited, as in fact they are. The two-stream theory also seems to account for many features of auroral radar echoes.[67,68] All in all, the two-stream theory seems to be clearly on the right track, although it leaves several effects still unexplained.

Two-stream instability theories for the electrojet have also been developed from the fluid equations.[69,70] For radar frequencies of 50 MHz, the use of the fluid equations is questionable, since k^{-1} is not very much larger than the ion mean free path (the values of collision frequency given

[66] D. T. Farley, *Phys. Rev. Lett.* **10**, 279 (1963); *J. Geophys. Res.* **68**, 6083 (1963).
[67] R. S. Unwin and F. B. Knox, *J. Atmos. Terr. Phys.* **30**, 25 (1968).
[68] R. L. Leadabrand, J. C. Schlobohm, and M. J. Baron, *J. Geophys. Res.* **70**, 4235 (1965).
[69] O. Buneman, *Phys. Rev. Lett.* **10**, 285 (1963).
[70] P. Waldteufel, *Ann. Geophys.* **21**, 579 (1965).

14.4. SCATTERING FROM THE EQUATORIAL ELECTROJET

in Table 1 must be reduced by roughly a factor of 3 since the center of the electrojet is about one scale height above 100 km). However, the agreement between the fluid equation and Vlasov equation results is quite good in general, with only minor numerical differences.

Although the two-stream theory provides some understanding of the electrojet phenomena, it is a highly oversimplified theory and leaves a number of questions unanswered. First, the theory is linearized and so can only predict the instability threshold; no quantitative prediction of the strength of the scatter can be given. Second, the medium is assumed to be homogeneous and infinite, even though the experimental data suggest that the gradients in collision frequency and electron drift velocity may be important. Among the more significant questions raised by the observations and not satisfactorily explained by the two-stream theory are the following:

1. Why are any echoes seen from overhead? The horizontal electron flow cannot couple to the vertically traveling waves.
2. What is the cause of the weak type II irregularities that appear before the two-stream threshold is reached?
3. What causes the irregularities to form in patches?
4. What is the nonlinear mechanism that limits the growth of the irregularities and determines the strength of the scattered signal?
5. What determines the complete shape of the spectrum? The two-stream theory accounts only for the peak in the spectrum of the strong echoes.

Dougherty and Farley[71] showed that some of these questions could perhaps be answered at least qualitatively by examining the nonlinear coupling of two of the unstable waves to form a third wave. Using only the coupling conditions

$$\omega_3 = \omega_1 \pm \omega_2, \quad \mathbf{k}_3 = \mathbf{k}_1 \pm \mathbf{k}_2, \quad (14.4.3)$$

it can be shown that the overhead echoes and many of the spectral shapes observed are at least possible. No detailed coupling mechanism was considered, however, and no quantitative calculations were made. Such nonlinear effects are probably important, but certainly it appears now that they cannot provide all the answers. In particular, they cannot account for the type II irregularities which appear before the two-stream threshold is reached. As Balsley[65] has pointed out, his experimental results suggest

---

[71] J. P. Dougherty and D. T. Farley, *J. Geophys. Res.* **72**, 895 (1967).

strongly that the shear in the electron drift velocity at the top and bottom of the current carrying layer must play a role in the formation of these irregularities, but as yet no theory has been developed.

Skadron and Weinstock[72] have attempted to show how the two-stream instability is stabilized by nonlinear effects. The "turbulent" electric field associated with the spectrum of unstable waves causes an enhanced diffusion, primarily of the ions, which finally stops the growth. Some of the assumptions made in the paper appear to be questionable, however, and some of the conclusions reached do not agree too well with the observations. Nevertheless, the paper is certainly an interesting first step in the direction of a satisfactory nonlinear theory.

14.4.5. Future Research

The linearized two-stream theory indicates that the peak in the scatter spectrum should occur at a Doppler shift which is almost, but not quite, proportional to the radar operating frequency. The phase velocity of the plasma waves and the threshold electron drift velocity for instability should increase slightly with decreasing wavelength. It is now possible to test these ideas by making simultaneous measurements at 16, 50, and 148 MHz at Jicamarca. A few preliminary measurements seem to be at least approximately consistent with the theory, but more data are needed.

Some attempts have been made to use higher order spectral functions, as discussed in Section 14.2.2.5, in the study of the electrojet scatter. Few results are presently available, but it does appear that these techniques may provide a powerful tool in the future for the study of nonlinear effects. One of the principal drawbacks at the moment is the high cost of the vast number of computations required.

Turning to theory, we must first of all explain the generation of the weak type II irregularities. Next we need acceptable nonlinear theories that give at least semiquantitative results for the amplitudes of both the strong and weak echoes.*

---

* ACKNOWLEDGMENTS: Part of this work was supported by the Arecibo Observatory, which is operated by Cornell University under contract to the National Science Foundation and with partial support from the Advanced Research Projects Agency. Support was also provided by the Atmospheric Sciences section of the National Science Foundation under NSF Grant GA–4350.

[72] G. Skadron and J. Weinstock, *J. Geophys. Res.* **74**, 5113 (1969).

15. DENSE PLASMA FOCUS*

15.1. Introduction

Dense plasma focus (DPF)[1-15] is a newly discovered plasma discharge with plasma densities $n > 10^{19}/\text{cm}^3$ and temperatures of a few kilo electron volts lasting 100 to 150 nsec. The focus is conjectured to be a short but finite two-dimensional z-pinch forming near or at the end of a coaxial plasma accelerator. Part of the stored magnetic energy in the tube and external circuit is rapidly converted to plasma energy during the current

[1] J. W. Mather, *Phys. Fluids Suppl.* **7**, 5–28 (1964).

[2] J. W. Mather, *Phys. Fluids* **8**, 366 (1965).

[3] P. P. Petrov, N. V. Filippov, T. I. Filippova, and V. A. Khrabrov, *in* "Plasma Physics and the Problems of Controlled Thermonuclear Reactions" (M. A. Leontovich, ed.), Vol. IV, p. 198. Pergamon, New York, 1960.

[4] N. V. Filippov, T. I. Filippova, and V. P. Vinogradov, *Nucl. Fusion Suppl. Pt. 2*, 577 (1962).

[5] N. V. Filippov and T. I. Filippova, *Plasma Phys. Contr. Nucl. Fusion Res. Proc. Conf., 2nd, 1965*, **2**. IAEA, 1966.

[6] J. W. Mather, *Plasma Phys. Contr. Nucl. Fusion Res. Proc. Conf., 2nd, 1965*, **2**. IAEA, Vienna, 1966.

[7] J. W. Mather and P. J. Bottoms, *Phys. Fluids* **11**, 611 (1968).

[8] E. L. Beckner, *J. Appl. Phys.* **37**, 4944 (1966).

[9] J. W. Long, N. J. Peacock, and P. D. Wilcox, *Proc. APS Topical Conf. Pulsed High-Density Plasmas, September 1967*, Los Alamos Sci. Lab. Rep., LA-3770, p. C5–1.

[10] D. A. Meskan, H. L. Van Paassen, and G. G. Comisar, *Proc. APS Topical Conf. Pulsed High-Density Plasmas, September 1967*, Los Alamos Sci. Lab. Rep., LA-3770, p. C6–1.

[11] N. V. Filippov, T. I. Filippova, V. I. Agafonov, G. V. Golub, L. G. Golubchikov, V. F. D'yachenko, V. D. Ivanov, V. S. Imshennik, YU. A. Kolesnikov, and E. B. Svirsky, *Plasma Phys. Contr. Nucl. Fusion Res. Proc. Conf., 3rd, Novosibirsk, USSR, 1968*, **2**. IAEA, Vienna, 1969.

[12] C. Maisonnier, F. Cipolla, C. Gourlan, M. Haegi, J. G. Linhart, A. Robouch, and M. Samuelli, *Plasma Phys. Contr. Nucl. Fusion Res. Proc. Conf., 3rd, Novosibirsk, USSR, 1968*, **2**. IAEA, Vienna, 1969.

[13] G. Decker, D. J. Mayhall, O. M. Friedrich, and A. A. Dougal, *Bull. Amer. Phys. Soc.* [2] **13**, 1543 (1968).

[14] D. L. Lafferty, D. C. Gates, and C. K. Hinrichs, *Bull. Amer. Phys. Soc.* [2] **13**, 1544 (1968).

[15] G. Patou, A. Simonett, and J. P. Watteau, *Proc. APS Topical Conf. Pulsed High-Density Plasmas, September 1967*, Los Alamos Sci. Lab. Rep., LA-3770, p. C2–1.

---

* Part 15 by J. W. Mather.

sheath's brief collapse toward the axis. The current-flow convergence appears to be largely due to the self-constricting nature of the current filament, whereas the heating and compression from the r, z implosion on the axis are due both to the magnetic forces of the current-carrying plasma filament and to the inertial forces. Partial conversion of the kinetic energy of the imploding axisymmetric current to internal heat energy may occur during the implosive phase owing to self-collision. With adiabatic compression, the final temperature can then be elevated rapidly.

The focus duration is many times greater than calculated from simple hydromagnetic theory. The $m = 0$ hydromagnetic instability leads to very rapid growth, $t = r_p/v \simeq 10^{-9}$ sec, where r_p and v are the final pinch radius and local ion speed, respectively. Several explanations attempt to justify the plasma longevity in terms of the **B** × ∇**B** drift velocity[6] and the two-dimensional curvature[16] of the current flow. Longmire[6,17] believes that the stabilization of the $m = 0$ mode is the result of high-velocity axial gas flow along the focused column. Since no stable equilibrium exists in the axial direction, either a gas flow or a "cold" electron flow may rapidly remove instability information from the focus. This belief has been made more quantitative by recent theoretical work by Friedman[18] which demonstrates stabilization as a result of longitudinal velocity gradients $\partial V_z/\partial z$ in the focus column.

The mechanisms operating in the focus region are still uncertain, although present neutron measurements[6,11,19] appear to favor a thermal moving plasma model in contrast to a high-voltage acceleration[10,12] (beam target) model. The scaling law for focus development favors an $I^2 \approx nKT$ dependence, i.e., pressure balance, whereas the scaling law for neutron production from several experiments[11,12,14,20] ranges from $W^{1.5}$ (W is the stored energy) to $W^{2.5}$. It is experimentally difficult to establish a simple neutron scaling law because the focus discharge operation depends not only upon the physical size of the tube but also upon the electrical and gas loading parameters.

The mechanisms for rapidly heating electrons are yet more uncertain. X-ray absorption measurements indicate electron temperatures of 1–3 keV ($Z = 1$). However, detection of high-Z X radiation in the focused region in some experiments casts suspicion on the exact interpretation or uni-

[16] G. G. Comisar, *Bull. Amer. Phys. Soc.* **13** [2], 1544 (1968).

[17] C. L. Longmire, Unpublished work. Los Alamos Sci. Lab. Los Alamos, New Mexico, 1967.

[18] A. M. Friedman, Unpublished work. Inst. of Nucl. Phys., Novosibirsk, USSR, 1968.

[19] P. J. Bottoms, J. P. Carpenter, J. W. Mather, K. D. Ware, and A. H. Williams, *Plasma Phys. Contr. Nucl. Fusion Res. Proc. Conf., 3rd, Novosibirsk, USSR, 1968*, **2**. IAEA, Vienna, 1969.

[20] N. J. Peacock, *Plasma Phys. Contr. Nucl. Fusion Res. Proc. Conf., 3rd, Novosibirsk USSR, 1968*, **2**. IAEA, Vienna, 1969.

15.1. INTRODUCTION

versality of such measurements. Electron heating can result from such secondary processes as ion–electron collisions, adiabatic compression and recompression, turbulent heating due to an "anomalous" resistance beam instability, and direct high-voltage acceleration. Most of these processes and others have been discussed by Tsytovich.[21] In fact, Tsytovich is concerned with the statistical acceleration of particles in plasma to explain both neutron and X-ray emission by high-power pulse discharges.[22] Some of the uncertainty about the X-ray emission processes in the focus could be reduced by measuring the intensity and electron temperature as a function of space and time.

The apparent focus-region stability depends upon the initial stored energy. X-ray pinhole-camera photography shows a definite tendency toward less stable focus development as the stored energy increases. Some recent work,[23] using small "axial" diverging magnetic fields ahead of the inner electrode, demonstrate greatly enhanced spatial stability of the focused column. With the magnetic field, the heating mechanism for ions and electrons is inhibited whereas the plasma volume is substantially increased. The plasma column resembles a "magnetic nozzle" diverging with distance from the inner electrode face. Axial magnetic field measurements near the plasma column are consistent with an azimuthally symmetric current sheath collapse.

Plasma acceleration by **j** × **B** forces has been studied experimentally and theoretically in many laboratories since the beginning of the controlled fusion program. The plasma focus discharge, originally studied by Filippov et al.[4] in the USSR and by Mather[1] in the USA around 1960, has attracted more interest since 1965, and now approximately 17 laboratories in six countries are engaged in focus research. The early work in the USA with coaxial gun systems was pioneered by Marshall.[24] The author,[1] following the work of Osher,[25] investigated the fast coaxial gun mode which led to the development of yet another, the high pressure (focus), mode of operation. Filippov et al., challenged by the problem of z-pinch initiation along insulators, designed a metal wall-pinch apparatus to minimize or eliminate the influence of insulating walls on the pinch. The Russian and United States' designs, although similar in many respects and results, were arrived at independently.

[21] V. N. Tsytovich, *Tr. P. N. Lebedev Phys. Inst.* **32**, 114 (1968).
[22] M. D. Razier and V. N. Tsytovich, *At. Energ.* **17**, 185 (1964).
[23] J. W. Mather, P. J. Bottoms, J. P. Carpenter, A. H. Williams, and K. D. Ware, *Phys. Fluids* **12**, 2343–2347 (1969); Los Alamos Sci. Lab. Rep. LA-4088-MS. Los Alamos Sci. Lab., Los Alamos, New Mexico, 1969.
[24] J. Marshall, *Phys. Fluids* **3**, 134 (1960).
[25] J. E. Osher, *Phys. Rev. Lett.* **8**, 305 (1962).

15.2. The Apparatus

Figure 1 is a typical schematic diagram of the coaxial plasma focus apparatus for a 27-kJ, 20-kV condenser power supply. The design shown in Fig. 2 is by Kolesnikov et al.[26] and is also used by Maisonnier et al.[12] The main design differences shown in Figs. 1 and 2 are in the relative electrode sizes and the absence in Fig. 2 of a current-sheath acceleration phase along the tube electrodes. In Fig. 2, the inner (1) and outer (3)

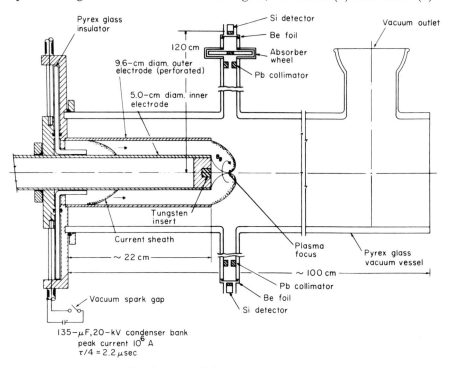

FIG. 1. A typical plasma focus apparatus.

electrode diameters are 48 and 70 cm, respectively, and the porcelain insulator (2) is 12-cm high. The axial spacing between anode and cathode is 8 cm. Item (7) of Fig. 2, is an azimuthal set of radial pins which are occasionally used to predetermine the filamentary structure of the sheath. The condenser supply (5) in this instance consists of 180 μF at voltages of 16–24 kV with a circular array of spark gaps (4). Filippov et al.[11] is presently operating a slightly larger 50-kV, 200-kJ focus apparatus.

[26] Iu. A. Kolesnikov, N. I. Filippov, and T. I. Filippova, *Proc. Int. Conf. Ioniz. Phenomena Gases, 7th, Belgrade, 1965*, **2**, p. 833.

15.2. THE APPARATUS

The design shown in Fig. 1 consists essentially of two coaxial copper electrodes and a Pyrex glass insulator across which the initial breakdown occurs. The outer electrode (OE) and center electrode (CE) diameters are typically 10 and 15 cm, respectively, with a CE length of 17 cm. For higher energy systems (67 kJ, 20 kV), OE and CE diameters of 15 and 10 cm, respectively, are employed with 25-cm electrode lengths. The minimum annular spacing, $\Delta r = 2.5$ cm leads to uniform breakdown at the insulator and symmetrical current sheath propagation. The OE is attached to a circular grounded cable header. The CE attaches to a central header that provides an electrical connection and vacuum seal. The hatlike Pyrex glass insulator forms the vacuum seal between the CE header and the grounded header. The central part of the insulator extends ~5 cm

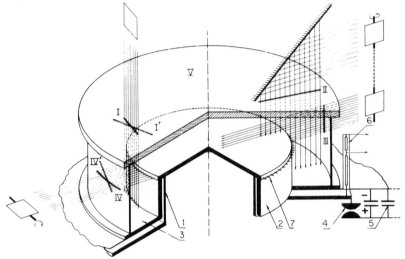

FIG. 2. Filippov's plasma focus apparatus. (Kurchatov Inst., Moscow, USSR.)

along the CE into the vacuum chamber. This insulator arrangement prescribes, by and large, the shape of the initial current sheath between the CE and the back plate of the OE. The exposed glass insulator surface is minimized by this geometry. Once gas breakdown is achieved at the insulator surface, the current buildup causes the current sheath to move radially outward in an inverse or unpinch manner, $j_z \times B_\theta$, forming a parabolic current sheath.

A vacuum vessel of Pyrex glass or copper surrounds the electrode structure. In early designs, Pyrex glass was used because it permitted photographing current sheath propagation along the tube. Today, especially at high electrical powers, metal chambers are used. Ordinary vacuum technology, utilizing oil-diffusion pumps and anticreep liquid N_2 traps, is employed. For high discharge pressures of 10–20 Torr, it is usually convenient to use additional cryosorb pumps filled with activated charcoal during the first minutes of recycling. In most DPF experiments today, the complete vacuum-chamber reactivation cycle is controlled by a sequential timer.

The condenser bank for the 27-kJ, 20-kV system consists of three condenser modules each comprising three 20-kV, 15-μF low inductance condensers connected in parallel. Each module is switched by means of 12 coaxial cables (Beldon YK 198, inductance $\sim$30 nH/ft) to the cable header by a centrally located vacuum spark gap.[27] The inductance of each modular unit, including the swtich, is $\sim$15 nH. The 67-kJ, 20-kV focus experiment, on the other hand, consists of eight such units; still higher energy systems may similarly be obtained. High-quality, long-lifetime, low-inductance condensers are readily available from such US manufacturers as Sangamo, Cornell–Dublier, and Aerovox.

The importance of low-inductance cables, condensers, and switches in the power supply cannot be overstressed if the large tube currents are to be achieved. With due consideration to the selection of low-inductance coaxial cable and condensers, the most important aspect of the power supply is the high-current, high-voltage switch. Many different switches of varying complexity exist today. Among these are the high-pressure nitrogen or sulfur hexafluoride gaps, dielectric switches, ignitrons, four-element electrode gaps using dry atmospheric nitrogen, and the graded vacuum spark gap.[27] The author has for many years been intensely interested in switches that can repeatedly transfer energies of 10–15 kJ. It is next to impossible to design a switch that will repeatedly carry unlimited quantities of current. Accepting this premise, the experimentalist is confronted with the problem of simultaneously closing many switches in parallel. Of the switches mentioned, the vacuum spark gap is, from the author's viewpoint, the most useful and reliable because of the following properties: (1) high current capability, $\sim$500 kA; (2) high voltage reliability, $\sim$50 kV; (3) triggering capability over a wide dynamic voltage range of 100 volts to 50 kV; (4) minimum time jitter of a few nanoseconds; (5) low inductance, 3–5 nH; (6) reasonably long lifetime, transferring $\sim$10 kJ for >4000 shots; (7) firing capability, depending on vacuum, of once per 60 sec (typical of LASL focus devices); and (8) triggering energy of $\sim$0.02 joules at 30 kV. With these switch properties, paralleling 100 to 200 10-kJ condenser modules to achieve 1- to 2-MJ energies becomes straight-forward. By comparison, the Scyllac θ-pinch condenser bank is designed to utilize 3200 switches for 10-MJ energy storage.

---

[27] J. W. Mather and A. H. Williams, *Rev. Sci. Instrum.* **31**, 297 (1960).

A cross section of the 20-kV vacuum spark gap and its coaxial condenser header[28] is shown in Fig. 3. The vacuum gap consists of three 1-cm-thick glass insulators and two floating brass electrodes held in compression by a nylon screw bushing. The space outside the glass insulators is pressurized with sulfur hexafloride to ~ 25-psi gage. The trigger plug is inserted centrally in the upper or ground (at $t = 0$) electrode. The trigger plug is of single- or two-electrode construction. The 20-kV switch is usually

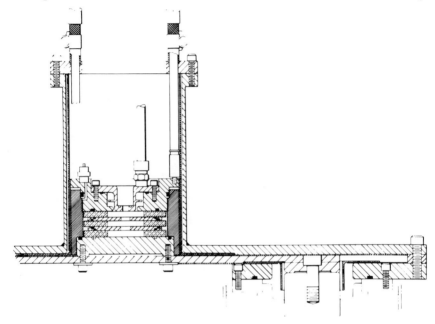

FIG. 3. The 20-kV, 9-kJ vacuum spark gap and condenser header.

"conditioned" under dc charging; more rapid conditioning for high-voltage operation can be achieved by pulse charging from a Marks circuit. Because of the importance of this process for successful vacuum switch operation, some explanation is required.

"Conditioning"[27] entails the formation of numerous metallic sites, none of which overlap, on the glass surface. These metallic sites become well dispersed over the surface and form a microscopically graded insulator across which large electric fields can be sustained without breakdown. The resistance of this glass-metal layer along the insulator surface is about 10^{12} to 10^{14} Ω. The conditioning begins with the development of fine

---

[28] A. H. Williams, K. D. Ware, J. W. Mather, J. P. Carpenter, and P. J. Bottoms, Bull. Amer. Phys. Soc. **13** [2], 1491 (1968).

surface cracks (crazing) on the glass surface caused by intense thermal heating from the first current discharge. Subsequently, the electrode metal vapor produced by the discharge deposits preferentially in these cracks. Once the atomic deposition begins, growth at these metallic sites proceeds readily, much like ordinary crystal growth. With the electron microscope, copper crystals have been seen in the original surface cracks; each metallic site is microscopically separated from its neighbor. The copper crystal growth is limited by subsequent current discharges which partially destroy the crystal buildup. The resulting surface, originally a poor heat conductor, develops a higher thermal conductivity which further limits the development of new surface cracks and new metallic sites.

15.3. Plasma Focus Development

The dense plasma focus accelerator requires, for optimum operation, simultaneous application of multiple condenser modules in ~ 10 nsec. The initial gas breakdown across the insulator and the development of the current sheath at the accelerator breech is not well understood. Experience shows that current sheath development requires prompt initiation; this promptness leads to a singular, azimuthally symmetric current front extending from the central positive electrode, back across the insulator surface, to the OE back plate.

Development of the plasma current leading to formation of a dense plasma region at the CE terminus can be conveniently subdivided into three main phases. First is the initial gas breakdown and the formation of a parabolic current front. Second is the hydromagnetic acceleration of a uniform axisymmetric current sheath toward the open end of the accelerator (this phase does not occur in the Filippov design). The third, and least understood, part of the discharge is the rapid collapse of the aximuthally symmetric current sheath toward the axis to form the dense plasma focus. Optimization of the last phase depends largely upon the magnetic energy stored in and outside the tube and upon the rate and manner with which it is converted to plasma energy.

15.3.1. Breakdown Phase

The early breakdown phase and its subsequent development was studied[29] with the aid of a fast image-converter camera. In Fig. 4 a series of axial photographs (5-nsec exposure) are shown as a function of time. Under final conditions that yielded good neutron production and soft X-ray pinhole photographs, it was found that gas breakdown occurred from

[29] J. W. Mather and A. H. Williams, *Phys. Fluids Res. Note* **9**, 2081 (1966).

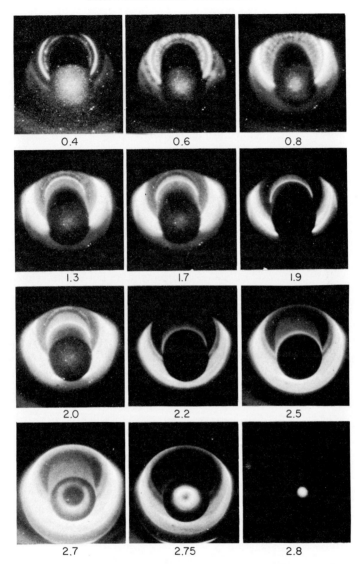

Fig. 4. Image-converter photographs (5-nsec exposure). Axial view of the early breakdown at the backplate leading to focus development (1 : 0.4).

the center positive electrode to the metal cathode back plate along the Pyrex glass insulator within a fraction of a microsecond after high-voltage application.

The breakdown in the first few tenths of a microsecond has a radial, striated light pattern with a definite multifilamentary structure. This structure, except for its obvious radial striation, appears cylindrically symmetric (note first few photographs in Fig. 4). As the tube current increases, the terminus of this visible pattern at the cathode back plate moves radially outward until it reaches the cylindrical part of the OE. The current front motion is best described as an unpinch or inverse process; i.e., the $(\mathbf{j} \times \mathbf{B})_r$ body force is exerted outward between the CE surface and the plasma current sheath. During this inverse phase, the sheath remains stable, as expected, because of the stability of the inverse pinch process (the B_θ lines are convex). The initial current-front shape is prescribed by the Pyrex insulator. This initial phase requires about 0.8 to 1.2 μsec, depending upon the initial gas loading and applied electric fields. In the final pattern recorded by the image-converter camera, all filamentary structures blend to form a uniform current discharge at ~ 1 μsec. The successful completion of this phase has been correlated with the final dense plasma formation and neutron production. Lafferty et al.[14] recently showed that when the plasma current sheath does not become azimuthally symmetric, but instead develops a preferred radial spoke, the neutron yield from a deuterium plasma focus is small or negligible.

In some recent work by Bostick et al.,[30-32] with a plasma-focus accelerator similar to that in Fig. 1 but used at much lower energies, a current-sheath filamentary pattern has also been observed with a fast image-converter camera. Although the early phases of this development resemble those shown in Fig. 4, these filamentary structures, rather than blending together, form a finite number of intense radial spokes which then propagate toward the open end of the coaxial accelerator. These so-called vorticity structures appear to retain their identity throughout the acceleration phase and finally coalesce or focus on axis beyond the end of the CE. There is strong evidence, according to Bostick, that these structures occur in pairs (force-free), and he states that the annihilation of their magnetic fields represents the main ion or electron heating mechanism (magnetic field annihilation) responsible for neutron and X-ray production.

Figures 5a and 5b are axial and transverse photographs, respectively, of typical filamentary structures of the current sheath during its convergence beyond the CE terminus. There is definite evidence in Fig. 5a that the radial structures occur in pairs although in many instances one or many

---

[30] W. H. Bostick, W. Prior, L. Grunberger, and G. Emmert, *Phys. Fluids* **9**, 2078 (1966).

[31] W. H. Bostick, G. Emmert, E. Farber, L. Grunberger, A. Jermakian, W. Prior, J. Zorskie, P. Casale, and A. Deshmukh, *Proc. Int. Conf. Ioniz. Phenomena Gases, 7th, Belgrade, 1965*, **2**, p. 141. Gradevinska Kryiga Publ., Belgrade, Yugoslavia.

[32] W. H. Bostick, L. Grunberger, and W. Prior, *Bull. Amer. Phys. Soc.* **13** [2], 1544 (1968).

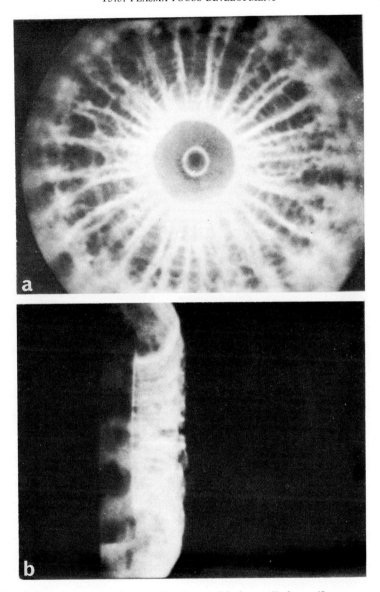

Fig. 5. Image-converter photographs of a coaxial plasma discharge (5-nsec exposure). (Stevens Inst. of Technol.) (1:1).

filaments appear in a given region. Whether or not these structures annihilate and give rise to neutron and X-ray production is, in the author's opinion, pure conjecture. These results were obtained with an 11-kV, 2.7-kJ coaxial discharge at $\sim$8-Torr deuterium pressure. Careful investigation[27] at higher energy levels failed to show such distinct features as those in Fig. 5. The coarser grain of this phenomenon at higher energies may be due to increased current-sheath plasma viscosity.

The importance of the initial breakdown and the subsequent development of an axisymmetric current sheath cannot be overemphasized. Quantitative evidence in terms of neutron and X-ray output shows that successful sheath generation depends on the applied electric field, gas pressure and kind of gas, contamination, electrode and insulator configuration at the breech, and the initial rate of change of current. Current carrier starvation during the few nanoseconds after breakdown may also dictate preferred sheath development.

15.3.2. Acceleration Phase

The description of the acceleration phase of the axisymmetric plasma sheath is almost classical. The current sheath across the annulus, Δr, is not planar, but canted backward from the anode to the cathode owing to the radial dependence of the magnetic pressure gradient. The total accelerating force, $\mathbf{j} \times \mathbf{B}$, acting perpendicular to the current boundary leads to radial and axial motion. The $(\mathbf{j} \times \mathbf{B})_r$ radial component is outward and, in a sense, forces the current sheath against the inner surface of the OE. The axial force $(\mathbf{j} \times \mathbf{B})_z$ varies as $1/r^2$ across the annulus and leads to higher sheath velocities near the surface of the CE. Magnetic probe measurements show that the current boundary velocity is higher near the CE than at the OE. Without the use of a perforated OE, the radial force term would lead to plasma pileup at the OE surface and in time would reduce the annular spacing, Δr. Owing to the parabolic current boundary, plasma flows centrifugally outward from anode to cathode along the current boundary as the acceleration continues. This centrifugal flow causes plasma stagnation and if not alleviated leads to a similar reduction in Δr. Fast image-converter photographs,[7] Fig. 6 (5-nsec exposure), taken as a function of time and transverse to the discharge through a perforated OE distinctly show a parabolic current front. Plasma is seen to progress radially outward and beyond the OE diameter as the current front accelerates downstream. Use of a perforated OE prevents plasma stagnation at the OE surface and substantially improves the quality of the focus and neutron production, indicating that electrode boundary conditions in the focus discharge are extremely important and must be included in any theoretical analysis.

The magnetic field distribution in the annular region was measured[2] as a function of r and z. Figure 7 shows the B_θ magnetic field as a function of z at a radius $r = 3.5$ cm at several times during the cycle. Most of the tube current is in a sheath approximately 1- to 2-cm thick as determined from $j_r \approx \Delta B_\theta / \Delta z$. The mass pickup by the current sheath was analyzed.

15.3. PLASMA FOCUS DEVELOPMENT

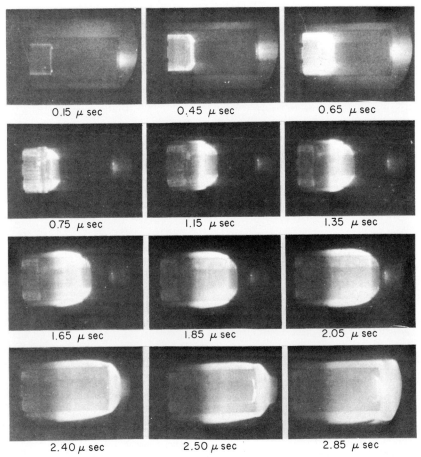

Fig. 6. Image-converter side-view photographs (5-nsec exposure) of the plasma focus discharge, taken through a perforated OE.

Assuming a priori that the current-layer velocity is the velocity of the instantaneous mass in the layer and also that the pressure gradient, ΔP, in the layer is negligible, one can equate the total rate of change of the layer momentum, ρv, to the **j** × **B** force. This is expressed as

$$\frac{\Delta \rho}{\Delta t} = \frac{(\mathbf{j} \times \mathbf{B})_z - \rho(dv_z/dt)}{v_z}. \tag{15.3.1}$$

The second term on the right-hand side is the ordinary (inertial) acceleration term, ρ is the instantaneous mass density, v the mass velocity, and **j** × **B** the $j_r \times B_\theta$ axial force component.

These results show that the instantaneous sheath mass, ρ_m, is almost a linear function of the axial distance for radial positions near the anode,

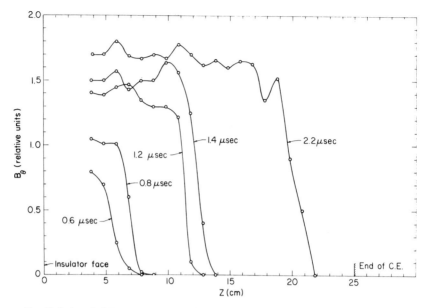

FIG. 7. Azimuthal B_θ magnetic field distribution as a function of z and time at a 3.5-cm radius for OE and CE diameters of 5 and 10 cm, respectively.

while the corresponding value of ρ_m near the cathode is extremely small. ρ_m at $r = 3.5$ cm increases by ~60%, whereas, at $r = 4.5$ cm, $\rho_m < 3\%$. Thus mass pickup is nonlinear with radius. This result, coupled with the magnetic pressure's radial dependence, can easily account for the time history, as well as the shape, of the current sheath with radius. The solution of Eq. (15.3.1) was made tractable because the average sheath velocity is constant at $\sim 10^7$ cm/sec. This result has been determined from fast photography, magnetic and electric probes, and the current and voltage waveforms.

The overall time for plasma sheath acceleration to the end of the CE is related to the applied voltage and the initial mass density of the static filling gas, according to the velocity relation determined by Rosenbluth and Garwin[33] for a model in which the current sheath "snowplows" the gas ahead of the sheath, as

$$v_s = (c^2 E^2/4\pi\rho_0)^{1/4} \quad \text{cm/sec}, \tag{15.3.2}$$

where E and ρ_0 are, respectively, the applied electric field and initial mass density (cgs) units.

The average sheath velocity has been obtained from the voltage and

[33] M. Rosenbluth and R. Garwin, Los Alamos Sci. Lab. Rep. LA-1850. Los Alamos Sci. Lab., Los Alamos, New Mexico, 1954.

current oscillogram by noting the time at which the singularity occurs in the waveforms. The singularity is characteristic of the occurrence of a rapid compression process which leads to the formation of a highly inductive current stream on-axis beyond the CE face. A typical current and

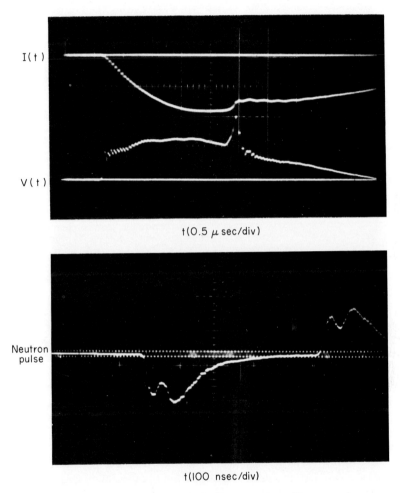

FIG. 8. Typical current (upper) and voltage (lower) oscillograms. Sweep speed, 0.5 μsec/div.

voltage waveform with this characteristic is shown in Fig. 8 (sweep speed 0.5 μsec/div). Figure 9 is a plot of this characteristic time as a function of applied voltage (pressure constant) or initial gas filling pressure (voltage constant). Good agreement with the prediction of Eq. (15.3.2) (solid

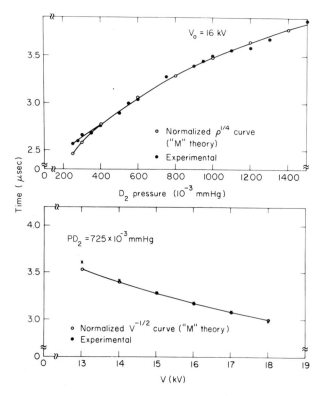

Fig. 9. Time to the current singularity as a function of applied voltage (pressure constant) and deuterium pressure (voltage constant).

curve) is obtained except for <400-μ deuterium pressures and for <13-kV voltages.

In the acceleration stage, the current sheath gains momentum as the tube current continues to increase. The current-sheath arrival at the end of the accelerator is experimentally adjusted to occur at peak tube current time. This is controlled, for a particular electrode geometry, by adjusting the CE length, the applied voltage, and the initial gas pressure. Different gases (H_2, D_2, He, Ne, and Ar) show that the mass density of the filling gas controls the sheath dynamics at a given voltage. Hence, the focus discharge can be used with different gases by simply reducing the initial particle density in the inverse proportion to the ion mass.

To optimize the conversion of stored magnetic energy to plasma energy during the third phase of the discharge, the dense plasma formation period, one must usually maximize the magnetic energy stored behind the current sheath in the accelerator just before its collapse to the axis. For a given external condenser circuit, maximum conversion of electrical capacitor energy to inductive magnetic energy occurs when the tube current is greatest. The time at which this occurs is governed largely by the flight time of the plasma sheath from the breech to the muzzle of the co-axial structure. Magnetic inductive storage can be sampled or used during a time governed by the speed of electromagnetic wave propagation along the coaxial tube, whereas the time to remove energy from a capacitor is dictated by the time $\tau \approx (LC)^{1/2}$, which is much too slow compared to experimental collapse and heating times of ~ 0.1 μsec. Time delay measurements, to be discussed later, suggest that the plasma sheath "snowplows" most of the gas it encounters during the acceleration phase. This means that the amount of ionized plasma left behind by the sheath is probably small.

15.3.3. Collapse Phase

The third phase encompasses the rapid convergence of the axisymmetric current sheath to the axis and the conversion of stored magnetic energy to plasma energy in the focus. The two-dimensional r, z convergence is due to the $\mathbf{j} \times \mathbf{B}$ pinch force. With this configuration, there is no equilibrium along the axis; hence, the plasma may readily escape axially in either direction. By the very nature of the convergence, much of the gas that the sheath encounters during collapse is ejected downstream and lost. The gas trapped in the focus is estimated[5] as $\sim 10\%$ of that originally there. It is this fact that leads to very highly compressed, small plasma.

The pinch effect is perhaps the most efficient way of heating and compressing a plasma, but, at the same time, it produces fast growing hydromagnetic instabilities. The implosion velocity of the current boundary imparts the same velocity to both ions and electrons, and because of the ion-electron mass difference, most of the energy appears as kinetic energy of the ions. Thus, in pinch devices, the ions are preferentially heated.

The sheath density during the radial collapse is substantially greater than ambient density, and the directed ion energy is rapidly randomized by self-collision; the same occurs for the electrons, but more rapidly. Hence, it is natural to think of a two-fluid plasma at different random temperatures. Average plasma-collapse speeds of approximately 4 to 6×10^7 cm/sec have been observed. From pinch theory calculation,[33] terminal ion speeds of about 7 to 9×10^7 cm/sec would be expected in the last moments of collapse.

Plasma heating and compression must be largely due to the magnetic and inertial forces during the on-axis current flow convergence. The inertial forces lead to high compression in addition to the magnetic forces in the current-carrying converged plasma filament. Investigators[5,6] of the focus have found that a small amount of krypton or xenon added to deuterium gas sometimes gives larger neutron production. To explain this result, Artsimovich,[34] in 1965, advanced the hypothesis that the deuteron energy may increase through collisions with heavy, high-energy ions, thereby increasing the deuteron temperature and neutron production. It is also *hypothesed* by others that the heavy ion concentration may enhance the inertial force through the mass enough to produce higher plasma compression. Each of these processes can lead to further heating and compression. Morozov's[35] theoretical work on steady-state flow shows that the inertial term in the theory leads to highly compressed plasma states of the on-axis flow.

15.3.4. Sheath Resistance

The current sheath resistance as a function of time is determined by a coaxial voltage probe placed on axis and downstream of the center electrode. The inner and outer conductor of the probe is attached to the face of the CE and OE of the accelerator, respectively. With the OE at ground potential, the probe measures the voltage between the coaxial electrodes during the acceleration and collapse of the discharge current. This voltage measurement is largely noninductive; i.e., the magnetic field, B_θ, ahead of the current sheath is negligible. Although one may argue that some B_θ field diffuses ahead of the resistive current sheath during acceleration, the time for field penetration can nevertheless be shown to be negligible during the time of interest. Knowing the tube current $I(t)$ and the interelectrode voltage, the sheath resistance R can be obtained as a function of time. Sheath potentials of about kT/e at the electrode surfaces have been ignored because they may not exist due to their rapid destruction by microinstabilities in the layer. A typical time-dependent resistance is shown in Fig. 10. The resistance decreases rapidly to a few milliohms in a few tenths of microseconds, then decreases slowly during the acceleration phase to ~ 50 $\mu\Omega$ at peak current time. The low plasma resistance at peak current time suggests a highly conductive current layer. According to the Spitzer resistivity relation,[36] the current sheath electron temperature is approximately 50 to 70 eV. During the collapse, the current sheath resistance increases slightly at first, then rises more rapidly reaching ~ 9 mΩ before the probe is destroyed. The resistance increase may be due to the rapid change of the focus geometry or to an "anomalous" plasma resistance caused by plasma turbulence. Although

[34] L. A. Artsimovich, Unpublished work. Kurchatov Inst., Moscow, USSR, 1965.

[35] A. I. Morozov, *Plasma Phys. Contr. Nucl. Fusion Res. Proc. Conf., 3rd, Novosibirsk, USSR, 1968*, 2. IAEA, Vienna, 1969.

[36] L. Spitzer, "Physics of Fully Ionized Gases." Wiley (Interscience), New York, 1956.

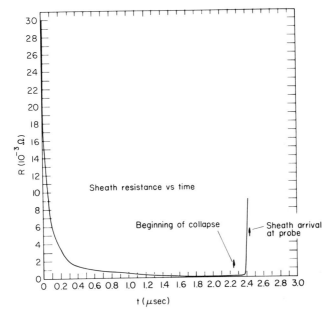

FIG. 10. Plasma current sheath resistance as a function of time.

the latter explanation may account for additional electron heating, some early results from a two-dimensional hydromagnetic code calculation have shown that the ordinary Spitzer resistivity can nevertheless account for the increased resistance during collapse.

15.3.5. Dynamic Behavior of the Current Sheath

The dynamic behavior[7,15] of the plasma current sheath has been determined using the measured time dependent values of the voltage $V(t)$ across the electrodes, the tube current $I(t)$, and the current sheath resistance $R(t)$.

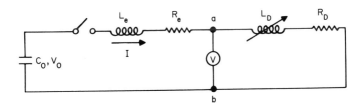

FIG. 11. Equivalent electrical circuit for the plasma focus discharge.

The circuit equation representing the voltage across the electrodes, ab, in the equivalent electrical circuit, Fig. 11, is

$$V(t) = (d/dt)[L_D(t)I(t)] + I(t)R_D(t). \tag{15.3.3}$$

The part to the left of ab in Fig. 11 represents the electrical storage capacitor and the leads; the part to the right, the discharge electrodes and the plasma system; $V(t)$ is the voltage across the electrodes, $L_D(t)$ is the discharge tube inductance, $I(t)$ is the current from the condenser bank, and $R_D(t)$ is the effective resistance (see Section 15.3.4) of the plasma current sheath as derived from V_r. The values of R_e (external circuit resistance) and L_e (external parasitic inductance) are determined when the electrodes, ab, are electrically shorted. Rationalized mks units are used throughout.

Using Eq. (15.3.3), the following quantities are calculated:

(a) Discharge inductance:

$$L_D(t) = \int_0^t [V(\lambda) - I(\lambda)R(\lambda)] \, d\lambda/I(t). \tag{15.3.4}$$

(b) Magnetic energy storage:

$$E_1(t) = \tfrac{1}{2}(L_e + L_D)I^2. \tag{15.3.5}$$

(c) Mechanical energy of the sheath, i.e., the sum of translational and thermal energy

$$E_0(t) = \tfrac{1}{2} \int_0^t \dot{L}_D I^2 \, d\lambda, \tag{15.3.6}$$

where $\dot{L}_D$ is obtained from Eq. (15.3.5).

(d) Pinch voltage during collapse:

$$V_p(t) = \dot{L}_D I + IR_D = V - L_D \dot{I} - IR_D. \tag{15.3.7}$$

The instantaneous mass of the plasma sheath can be estimated using the momentum equation, $d(\rho v)/dt = \mathbf{j} \times \mathbf{B}$. Assuming that the azimuthal field $B_\theta \sim 1/r$ and $dv/dt \simeq 0$ (this is justified from the slope of $L_D(t)$ up to the moment of collapse), one obtains

$$m_0(t) = [10^{-7} \int_0^{t_c} I^2 \ln(r_0/r_i) \, d\lambda]/v, \tag{15.3.8}$$

where v is the axial sheath velocity along the tube and r_0/r_i is the ratio of the outer to inner electrode radii. The upper integration limit, t_c, is a cutoff time corresponding to the singularity in the current oscillogram, Fig. 8, which represents the time when the fast radial collapse begins. For times $t > t_c$, the actual velocity function, $v(r, z)$, is unknown since the

15.3. PLASMA FOCUS DEVELOPMENT

collapse is actually a two-dimensional compression. Once $m_0(t)$ is known, the sheath kinetic energy, $m_0 v^2/2$, is obtained and the internal energy per unit mass may then be computed from Eq. (15.3.6) as

$$E(t) = \tfrac{1}{2}(\{\int_0^{t_c} I^2 \dot{L}_D \, d\lambda\}/m_0 - v^2). \tag{15.3.9}$$

To a good approximation, v is a constant ($\sim 10^5$ meter/sec). This has been verified with magnetic probes and fast photography; it is now obtained from the time derivative of Eq. (15.3.4). For a coaxial system of fixed ratio r_0/r_i, the average value of the rate of change of inductance depends only on the sheath velocity, v.

Equations (15.3.4) through (15.3.9) were coded[37] for the 7030 computer, and a typical set of results is shown in Fig. 12. Curves (a), (b), and (c) represent the input current $I(t)$, voltage $V(t)$, and plasma sheath resistance $R_D(t)$, respectively. Curves (d)–(i) represent the quantities given by Eqs. (15.3.4) through (15.3.9), respectively. These results are for the following initial conditions: $V_0 = 17.3$ kV, $C = 90$ μF, $L_e = 16$ nH, $R_e = 3.3 \times 10^{-3}$ Ω, ρ_0 (initial deuterium density) $= 4.9 \times 10^{17}$ particles/cm^3, $r_0/r_i = 2$, and the accelerator length is 0.2 meters. Peak current is 537 kA, and the time, t_c, to the current singularity is 2.12 μsec.

The following remarks can be made:

1. The discharge inductance increases linearly with time; for $t > t_c$, L_D rises rapidly during the fast radial collapse. The instantaneous value of L_D at $t_c = 2.12$ μsec is 30.5 nH; the subsequent increase corresponds to an average rate of change of inductance of 0.05 H/sec.

2. $E_1(t)$ rises from zero to 6.68 kJ at $t_c = 2.12$ μsec, then decreases during the collapse; in this case the average power supplied to the collapse stage is $\sim 7 \times 10^9$ joules/sec. The energy going into the collapse stage comes in part from the magnetic energy stored behind the current sheath both in the discharge tube and in the external parasitic inductance. This result agrees with the operation principle stated earlier, that energy is stored inductively behind the moving current sheath and then rapidly converted to plasma energy during the formation of the dense plasma focus.

3. The mechanical sheath energy $E_0(t)$ increases in a manner similar to $E_1(t)$ up to $t_c = 2.12$ μsec; for $t > t_c$, $E_0(t)$ rises rapidly to ~ 3.2 kJ; this energy is plasma kinetic and internal energy.

4. The voltage across the moving plasma, $\dot{L}_D I$, increases with time and reaches ~ 8 kV at $t = t_c$; during collapse ($t > t_c$), the voltage rises to ~ 40 kV primarily because of the large inductive change of the current

---

[37] C. Longmire and J. Longley, Unpublished work. Los Alamos Sci. Lab., Los Alamos, New Mexico, 1966.

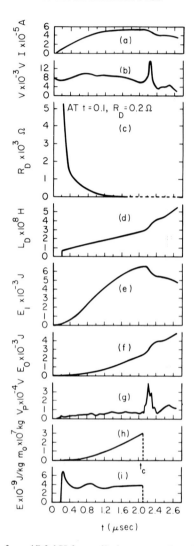

FIG. 12. Typical data for a 17.3-kV focus discharge as a function of time. (a) current, $I(t)$; (b) voltage across ab, $V(t)$; (c) sheath resistance, $R_D(t)$; (d) sheath inductance, $L_D(t)$; (e) inductive magnetic energy, $E_1(t)$; (f) mechanical energy of sheath, $E_0(t)$; (g) pinch voltage, $V_P(t)$; (h) plasma sheath mass, $m_0(t)$; and (i) internal energy of plasma sheath per unit mass, $E(t)$. Curves (a)–(c) represent measured quantities; mks units are used.

sheath, and then drops to ~7 to 8 kV at the end of the collapse. For higher energy focus experiments (80 kJ), V_p rises to 80–100 kV. The full-width, half-maximum time duration of the generated pinch voltage is ~0.12 μsec.

15.3.6. Energy Partition in the Discharge

The data shown in Fig. 12 can be used to determine the partition of the initial condenser energy in the discharge system. Curves e and f are the variable magnetic storage, $E_1(t)$, and the mechanical sheath energy, $E_0(t)$, respectively. Curve c is the variable plasma resistance, $R_D(t)$; the value of $R_D(t)$ for $t > t_c$ (shown dotted) is assumed constant for this analysis. The external resistance, R_e, is $\sim 3.3 \times 10^{-3}$ Ω. At peak current time, when $dI/dt = 0$, the voltage at the electrical storage condenser terminals becomes $(R_e + R_D + \dot{L}_D)I$. This value is proportional to the charge left on the capacitors and, hence, represents ~ 2.7 kJ of unused electrical energy.

At $t = t_c = 2.12$ μsec, energy is partitioned as follows: $E_1 = 6.68$ kJ, and $E_0 = 2.34$ kJ and the external and plasma-integrated ohmic losses, I^2Rt, of 1.3 and 0.3 kJ, respectively. The sum of these energies agrees with the initial electrical storage, 13.4 kJ, minus the unused portion, 2.7 kJ.

For times $t > t_c$, one can estimate how the energy may partition during the collapse time, the time between $t = t_c$ and the moment of final collapse. From Fig. 12e, one sees that only 1 kJ of the stored magnetic energy $LI^2/2$ is converted during ~ 0.2 μsec. However, Fig. 12f shows that mechanical energy of the collapsing sheath increases by ~ 1.2 kJ during the same time. In addition to the 1.2 kJ, energy must be supplied to offset the ohmic losses of the external circuit resistance (~ 0.17 kJ) and the increased plasma resistance. Thus the total energy associated directly or indirectly with the collapse process is at least 1.37 kJ; this value is >1 kJ as indicated by direct conversion from Fig. 12e. To account for this apparent discrepancy, one must consider the residual energy (2.7 kJ) stored in the electrical capacitor at the moment of collapse; i.e., all the condenser energy had not been delivered to the discharge by peak current time. Of the 2.7 kJ, an appreciable part can be delivered rapidly to the discharge circuit. For example, the maximum rate of transfer of electrical energy to inductive energy from the condenser is 0.95 kJ/0.2 μsec. Thus, the circuit can transfer ~ 0.95 kJ of the 2.7 kJ (residual capacitor energy storage) during collapse. The ambiguity in the energy balance can now be resolved by noting that 1 kJ of direct inductive and ~ 0.95 kJ of direct capacitor-coupled energy can be utilized. This corresponds to a maximum energy conversion of ~ 1.95 kJ in ~ 0.2 μsec, while the energy required in the collapse stage is ≈ 1.4 kJ. Unquestionably, a large part of the mechanical energy of sheath motion is lost in terms of plasma ejection. This is presumably a consequence of the noncylindrical (r, z) collapse. Of the 1.2 kJ of mechanical energy in sheath motion only a fraction ($\sim 25\%$) is recovered as plasma energy in the focus.

The fractional conversion of stored magnetic energy behind the moving plasma sheath to plasma energy during the plasma collapse can be related to the current change at peak current time. Assuming that the magnetic flux, LI, remains constant during the rapid collapse stage, and neglecting the ohmic energy loss term of the collapse in the energy conservation equation, one obtains for the minimum magnetic energy converted to plasma energy, W_p,

$$W_p = \tfrac{1}{2} L_i I_i^2 (1 - \alpha), \qquad (15.3.10)$$

where

$$\alpha \equiv I_f/I_i,$$

where I_f and I_i are the final and initial currents, respectively, of the collapse process.

For the specific case in Fig. 12a, $\alpha = 0.855$, E_1 ($t = 2.12$ μsec) $= 6.68$ kJ, and $W_p = \sim 0.99$ kJ. This is similar to the 1-kJ value read directly from Fig. 12e. The quantity α can be defined also as a geometrical parameter involving the external inductance, L_e, discharge length, l, and radius ratio r_o/r_i, much akin to the formulation of the ordinary pinch theory[33]; however, these dependences have not been checked experimentally, except to note that W_p is larger for smaller α.

The internal NkT energy of the compressed plasma pinch is estimated from the final discharge current and the length of the discharge column. For a pinch current I, the magnetic pressure, $B^2/2\mu_0$, acting over a pinch length l produces a constant NkT per unit length according to the Bennett[38] pinch relation, $I^2 = 2 \times 10^7 NkT$ (N is the final line density in particles per meter, and $T = T_e + T_i$). For $I \approx 459$ kA, the energy per unit length, NkT, is 10 kJ/meter. The total energy contained in a 0.03-meter pinched plasma at 2.31 μsec is ~ 300 joules. The time dependence of the current after collapse affects the focus quality. In some cases the current continues to decrease in time after the pinch formation, which in effect rapidly reduces NkT and decreases neutron production. In other instances, the current is fairly constant for several tenths of microseconds, and NkT is maintained; in these instances, neutron production is appreciably higher.

15.3.7. Impedance Considerations

In an LC circuit with a time-changing inductance, $\dot{L}$, the fraction of the initially stored electrical energy, $\tfrac{1}{2} C V_0^2$, converted to magnetic energy, $\tfrac{1}{2} L I^2$, is a function of the dimensionless parameters $\dot{L}/R_c$, where $R_c \equiv 2(L_e/C)^{1/2}$ is the external surge impedance. In the plasma focus experiment, the maximum available current is also of utmost importance because the final plasma pressure depends upon I^2. In the following

[38] W. H. Bennett, *Phys. Rev.* **45**, 890 (1934).

15.3. PLASMA FOCUS DEVELOPMENT

analysis,[39] the value of $\dot{L}/R_c$ determines both the available tube current and the fractional energy converted to magnetic energy.

The circuit equation for the equivalent circuit shown in Fig. 11 is solved in terms of the current $I(t)$ and its dependence on the electrical circuit parameters, L, $\dot{L}$, C, and R; $\dot{L}_D$ and $(R_e + R_D)$ in Fig. 11 are replaced by $\dot{L}$ and R, respectively. Equation (5.3.11) is rewritten as

$$(L_e + \dot{L}t)(dI/dt) + I(R + \dot{L}) + \int_0^t I/C\, dt = V_0, \quad (15.3.11)$$

with the initial condition $I(0) = 0$. The solution of Eq. (15.3.11) in terms of Bessel functions is

$$\frac{I(t)}{I_m} = \left(\frac{\alpha}{\omega}\right)^k \frac{J_k(\alpha)J_{-k}(\omega) - J_{-k}(\alpha)J_k(\omega)}{J_{k+1}(\alpha)J_{-k}(\alpha) + J_{-k}(\alpha)J_k(\alpha)}, \quad (15.3.12)$$

where

$$k = 1 + R/\dot{L}, \quad \alpha = R_c/\dot{L}; \quad R_c = 2(L_e/C)^{1/2},$$
$$\omega = (\alpha(\alpha + 4\pi t/\tau))^{1/2}, \quad \tau = 2\pi(L_e C)^{1/2}, \quad \text{and} \quad I_m = V_0(C/L_e)^{1/2}.$$

If k is an integer, the Bessel functions with negative indices become Bessel functions of the second kind. The current ratio, $I(t)/I_m$, as a function of the dimensionless time, t/τ, is plotted in Fig. 13 for various values of the dimensionless parameter $\dot{L}/R_c$ for the case of $R = 0$. Both the damping and the increasing period due to the changing inductance are apparent.

To maximize the available current for a given $\dot{L}$ term, R_c should be maximized. However, the value of R_c is not completely undetermined, in fact, R_c is, in a sense, predetermined by the capacitance, C, to achieve a given energy ($\frac{1}{2}CV_0^2$). On the other hand, the external parasitic inductance, L_e, is usually made as small as possible to obtain high rates of current change and large currents. This combination of small L_e and large C make R_c small, which is usually undesirable for a constant $\dot{L}$ load. A time-changing inductance, $\dot{L}$, in a high capacitance system leads to large $\dot{L}/R_c$ which markedly affects the available current, $I(t)$ (see Fig. 13).

The peak fractional magnetic energy stored internally between the coaxial electrodes, $W_I/W_0 = \frac{1}{2}LI^2/\frac{1}{2}CV_0^2$, is plotted versus the parameter $\dot{L}/R_c$ in Fig. 14. The peak quantity $W_T/W_0 = \frac{1}{2}(L_e + L)I^2/\frac{1}{2}CV_0^2$ is also shown; and represents the fraction of condenser energy storage appearing as total magnetic energy. It is seen that W_I/W_0 rises rapidly with $\dot{L}/R_c$ to a maximum of ~ 0.43 at $\dot{L}/R_c \simeq 0.7$ and then decreases slowly for larger $\dot{L}/R_c$ values. W_T/W_0, on the other hand, decreases steadily as $\dot{L}/R_c$ increases. For a coaxial electrode geometry, $\dot{L}$ is related

---

[39] D. Baker, Unpublished work. Los Alamos Sci. Lab., Los Alamos, New Mexico, 1967.

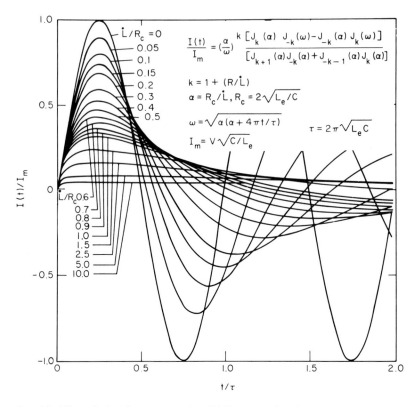

FIG. 13. The calculated current ratio, $I(t)/I_m$, as a function of the dimensionless parameter t/τ for a several values of the dimensionless parameter $\dot{L}R_c$ and $R = 0$. The influence of a time-changing inductance, $\dot{L}$, in an LC circuit reduces the available current, I.

to the electrode aspect ratio, r_0/r_i, and the sheath velocity, v_s, as $L = (\mu_0/2\pi)v_s \ln(r_0/r_i)$. Since the sheath velocity is apparently not a sensitive function of the electrical parameters, i.e., $v_s \approx$ const., $\dot{L}$ can, in principle, be reduced by decreasing r_0/r_i. Figure 15 shows the times, in units of $\tau = 2\pi(L_e C)^{1/2}$, at which the peak magnetic storage occurs. If the assumptions of constant sheath velocity v_s and $R \approx 0$ are valid, the maximum stored internal energy, W_1, occurs when the experimental parameters are adjusted so that $\dot{L}$ is ~ 0.7, the critical surge impedance of the external circuit. On the other hand, the quantity W_T has no definite maximum, as expected.

The above discussion is applied to the dense plasma focus accelerator, DPF II. For $C = 135$ μF and $L_e = 10$ nH, R_c becomes 0.017 Ω. The rate of change of the tube inductance $\dot{L}$, for an electrode aspect ratio $r_0/r_i = 2$

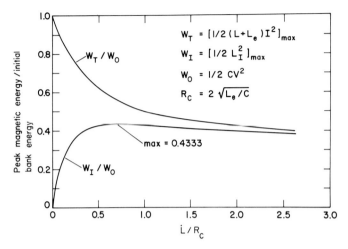

Fig. 14. Peak fractional magnetic energy, $W_I/W_0 = \frac{1}{2}LI^2/\frac{1}{2}CV_0^2$ and $W_T/W_0 = \frac{1}{2}(L_e + L)I^2/\frac{1}{2}CV_0^2$, versus the impedance-matching parameter $\dot{L}/R_c$.

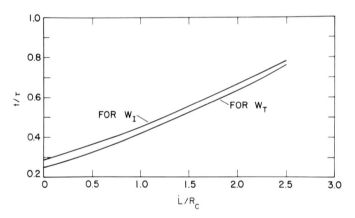

Fig. 15. Time t/τ [in units of $\tau = 2\pi(L_e/C)^{1/2}$] at which the peak magnetic storage, W_I and W_T, occurs versus $\dot{L}/R_c$.

and a sheath velocity of 10^5 meter/sec, is 0.014 Ω. Thus $\dot{L}/R_c \simeq 0.8$. From Fig. 14, an $\dot{L}/R_c = 0.8$ corresponds to a value of 0.41 for W_I/W_0 and ~0.54 for W_T/W_0. Certainly, on the basis of W_I/W_0, little is to be gained by juggling $\dot{L}$ or L_e; however, W_T/W_0 is quite sensitive to $\dot{L}/R_c$. Let us compute how far the current sheath has traveled at the time, t_m, at which $W_T = \frac{1}{2}(L_e + L)I^2$ is a maximum. From Fig. 15, this occurs at $t/\tau = 0.27$; $\tau = 7.3$ μsec so that $t_m = 2.7$ μsec. Subtracting from t_m a time required for the development of the inverse pinch phase (~1 μsec), the current sheath has traveled ~17 cm at the time W_T is a maximum ($v_s \simeq 10$ cm/μsec). This result agrees well with the actual 17.5-cm electrode length. The above analysis is useful in understanding the accelerator; however, other questions, such as how the plasma collapse utilizes the stored magnetic energy, and details of the pinch dynamics, and the heating mechanisms, need answering.

The importance of W_T or W_I, or both, depends upon whether the energy sink represented by the focus can communicate to all regions of stored inductive energy during collapse. The communication time is related to the magnetic field and the amount of ionized gas left behind by the current sheath. The residual ionized gas coupled with the magnetic field is a dielectric medium in which the propagation velocity of an electromagnetic wave is given by $v_w = c/k^{1/2}$, where c is the velocity of light and $k = 1 + 4\pi\rho c^2/B^2$ is the dielectric constant of the magnetized medium. If the communication time $t \equiv 1/v_w \gg t_c$, where 1 is the electrode length and t_c is the time of collapse to the focus, then only a fraction of that energy stored internally, W_I, can be utilized. For $t \ll t_c$, the reverse is true; the plasma collapse can sample the total stored magnetic energy, W_T, during the collapse phase. The propagation speed of an electromagnetic wave down the coaxial tube was obtained by determining the time delay between a dI/dt signal near the focus and its appearance at the breech of the accelerator. This time difference corresponds to a wave speed $v_w \simeq 10^9$ cm/sec[1]. On this basis, the energy communication time, t, between the focus region and external regions of stored magnetic energy is $\simeq 25$ nsec, much less than the duration of the collapse process, $t_c \simeq 150$ nsec. In this respect, the total magnetic energy, $W_T = \frac{1}{2}(L_e + L)I^2$, can be sampled during the current collapse. This result agrees well with the results of the dynamic behavior analysis in Section 15.3.5 and the calculated minimum energy conversion to plasma energy in Section 15.3.6 in which the total stored magnetic energy was used. Further confirmation of the wave speed was obtained from the time delay between an external dI/dt probe signal and the emission of hard X rays from the anode (the X-ray signal signifies the start of the collapse process). The time-delay measurements[40] can be taken one step further to estimate the average residual plasma density in the tube. For $v_w \approx 10^9$ cm/sec[1], a 25-cm-long coaxial tube, and an average magnetic field $B_\theta = 45$ kg, one obtains from the dielectric expression an average residual plasma mass density $\rho_r = 1.7 \times 10^{-10}$ gm/cm^3. This value is $<10^{-4}$ of the original mass density, $\rho_0 = 2 \times 10^{-6}$ gm/cm^3, in the tube which says, in effect, that the current sheath is an extremely efficient

[40] S. Glasstone and R. H. Lovberg, "Controlled Thermonuclear Reactions." Van Nostrand, Princeton, New Jersey, 1960.

"snowplow." The wave speed, v_w, in this case is also the Alfvén velocity when unity is neglected in the dielectric constant expression. A similar estimate of the residual density by Marshall (see Butler et al.[41]) yields a value of $\sim 1\%$.

15.4. Plasma Diagnostic Measurements

15.4.1. Neutron Production

The spectral and intensity distributions of D–D neutrons from high temperature plasma have been treated theoretically by several authors.[42,43] When the distributions are anisotropic, one can use the anisotropy diagnostically to determine the velocity distribution of the ions in the plasma. Neutron production from the dense plasma focus (DPF) has been found to be isotropic to within 5% in a double-ended device[2,11]; however, results[11] from the single open-ended device (Fig. 1) show definite anisotropy both in neutron energy and intensity. The measured angular distribution of neutron intensity and neutron energy is compared with proposed theoretical models of the plasma ion distribution. The neutron energy first observed is that of the fastest neutron obtained from the rapid rise of the neutron pulse. The neutron spectrum $f(E, t)$ is a convolution of two independent functions, one the energy spectrum and the other the time of production, so it is difficult to separate these quantities. To determine the time dependent neutron energy spectrum $f(E, t)$ during a single pulse would require the use of a proton magnetic spectrometer and neutron intensities $> 10^{11}$ neutrons/pulse.

The plasma focus apparatus used for the measurement to be described here (20-kV, 67-kJ) is similar to that in Fig. 1. The intensity measurements were obtained with a calibrated activated silver Geiger tube array housed in a $30 \times 30 \times 15$ cm cadmium-lined paraffin block.[44] A time-of-flight telescope was constructed using fast Amperex photomultipliers[45] and plastic scintillators. The response time of the scintillator is in the 1- to 2-nsec region. The Amperex 56 TVP photomultiplier circuit was designed for up to 1-ampere linear output into a 50-Ω load for pulse widths of up to 2-μsec duration. The photomultiplier rise time was ~ 2 nsec. Each

---

[41] T. D. Butler, I. Henins, F. C. Jahoda, J. Marshall, and R. L. Morse, Los Alamos Sci. Lab. Rep. LA-DC 9003. Los Alamos Sci. Lab., New Mexico, 1968.

[42] G. Lehner and F. Pohl, Z. Phys. **207**, 83 (1967).

[43] C. L. Critchfield, Unpublished work. Los Alamos Sci. Lab., Los Alamos, New Mexico, 1967.

[44] R. J. Lanter and D. E. Bannerman, Los Alamos Sci. Lab. Rep. LA-3498-MS. Los Alamos Sci. Lab., Los Alamos, New Mexico, 1966.

[45] L. K. Neher, Unpublished work. Los Alamos Sci. Lab., Los Alamos, New Mexico, 1967.

multiplier was housed in a double electrostatic shield to eliminate electrical noise pickup. The pulse data were displayed directly on the plates of a Tektronix 517 oscilloscope with 100-MHz response. An artificial delay line was inserted in the circuit of the farthest detector (flight path 3 meters) to achieve good pulse separation.

Data were obtained from many shots at observation angles 0°, 45°, 90°, and 180°. The results for the average highest neutron energies are given in Table I, column 2. These data show anisotropy in the neutron energy. As will be discussed shortly, these data are interpreted in terms of an isotropic plasma ion distribution in the center-of-mass system with the addition of a center-of-mass velocity in the axial direction.

The experimental accuracy of the velocity measurement is $\sim 1.5\%$, whereas the percentage of scatter in neutron energy on a pulse basis, from the average value is shown in Table I. In spite of the neutron-energy scatter from shot to shot, there is no correlation of the neutron energy with absolute neutron yield. We know from soft X-ray pinhole photographs (Section 15.4.4) that the plasma focus is not exactly reproduced spatially from pulse to pulse. So, to account for the above result, it is possible that the nonsymmetry or nonreproducibility in space may be augmented by an asymmetry in $v(r, z)$ in the last moments of collapse, causing the earliest neutron emission to exhibit a statistical variation in energy.

The neutron intensity distributions were also obtained at angles of 0°, 45°, 90°, and 180°. To make a first-order neutron-scattering correction to the data, a steady source of neutrons was placed at the plasma focus. Most of the background scattering material was high-Z; a very small part of the experimental area contained hydrogenous material. Nevertheless, small thicknesses of metal requires significant intensity corrections. The use of steady-state neutron sources, rather than pulsed sources, for calibration is certainly undesirable because of equilibrium buildup associated with the scattering material. This calibration problem confronts many investigators, and is not easily solved unless a pulsed neutron source is available. The angular distribution of the neutron intensity at 0° to the intensity at angle φ is shown in Fig. 16 for the 67-kJ device; $\varphi = 0°$ and $\varphi = 180°$ corresponds to the direction downstream and upstream respectively of the center electrode. The intensity ratio increases smoothly and slowly from 0° to 180°. The anistropy of the neutron intensity ratio is $\sim 5\%$ at 45°, $\sim 12\%$ at 90°, and $\sim 27\%$ at 180°. The experimental accuracy for $I(0)/I(\varphi)$ is $\sim 2.5\%$. The remaining curves in the figure pertain to theoretical models, and will be discussed.

Since the observed neutron distribution is anisotropic, one is led to consider plasma models in which a center-of-mass motion exists. Motion of the center of mass (c.m.) can arise in the accelerated or target model and

TABLE I

E_n (MeV) φ	Exp.	Moving isotropic dist., $kT = 0$	Monoenergetic beam[a]			Moving isotropic dist., $kT = 0$	Exp. $I(0)/I(\varphi)$
			$\theta_0 = 0°$	$\theta_0 = 30°$	$\theta_0 = 45°$		
0	2.77 ± 0.12	2.77	2.77	2.77	2.77	2.89	1
π/4	2.67 ± 0.15	2.68	2.68	2.80	2.86	2.8	1.047 ± 0.03
π/2	2.59 ± 0.09	2.47	2.47	2.66	2.77	2.59	1.119 ± 0.03
π	2.22 ± 0.12	2.2	2.2	2.27	2.35	2.31	1.268 ± 0.03

[a] ρ based on the experimental $I(0)/I(\pi/2)$ ratio

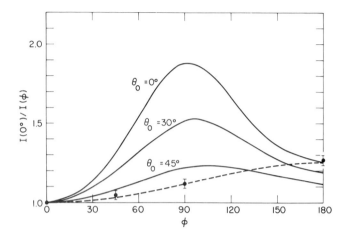

FIG. 16. The angular neutron-intensity distribution (normalized) (solid line) expected for the target beam model for several incident angles θ_0 and for the moving isotropic plasma model (dashed line). The experimental distribution is shown as dots (●).

in a model in which part or all of the plasma ions are in motion with $v_{\text{c.m.}}$ along the axis.

For a thermal plasma moving at velocity v, shown schematically in Fig. 17, the angular distribution is isotropic in the c.m. system. A neutron

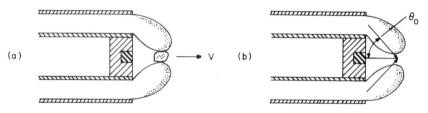

FIG. 17. (a) The moving isotropic plasma model and (b) the accelerated beam (target) model at angle θ_0 (b).

emitted with velocity u at angle θ in the c.m. relative to v will have a laboratory energy given by

$$E_n = \tfrac{1}{2}M(\mathbf{u} + \mathbf{v})^2, \tag{15.4.1}$$

where $\tfrac{1}{2}Mu^2 = \tfrac{3}{4}[Q + f(kT)]$ is the neutron energy in the c.m. system, M is the neutron mass, and $Q = 3.267$ MeV. Transforming θ to the laboratory angle φ gives

$$\cos \theta = (1 - \rho^2 \sin^2 \varphi)^{1/2} \cos \varphi - \rho^2 \sin^2 \varphi, \tag{15.4.2}$$

where $\rho \equiv v/u$.

Equation (15.4.1) is rewritten to give the neutron energy ratio at $\varphi = 0°$ relative to that at angles φ in the laboratory,

$$E_n(0)/E(\varphi) = (1 + \rho)^2/(1 + 2\rho \cos\theta + \rho^2), \quad (15.4.3)$$

and, similarly, the angular distribution of neutron intensity ratio in the laboratory system is given in terms of the cross section, $\sigma_0(\varphi)$, as

$$\sigma_0(0)/\sigma_0(\varphi) = (1 + \rho)^2(1 + \rho \cos\theta)/(1 + 2\rho \cos\theta + \rho^2)^{3/2}. \quad (15.4.4)$$

Equation (15.4.3) is valid for $kT = 0$, whereas Eq. (15.4.4) is independent of kT.

Now consider an accelerated deuteron beam model, Fig. 17b, in which the deuterons impinge upon a stationary gas target at an angle θ_0 to the axis and at all aximuthal angles ψ about it, keeping $\rho = v/u$ where v is the c.m. velocity. For the special case $\theta_0 = \theta$, the neutron energy angular distribution is the same as in Eq. (15.4.1), but the c.m. neutron energy, $\frac{1}{2}Mu^2$, is larger because the laboratory deuteron energy is nearly four times larger than the deuteron in the isotropic moving plasma ($kT = 0$); i.e., $\frac{1}{2}Mu^2 = \frac{3}{4}(Q + \frac{1}{2}E_d)$. In the more general θ_0 case, $\theta_0 \neq 0$, the laboratory neutron energy distribution is given approximately as

$$E_n = \tfrac{1}{2}M(\mathbf{u} + \mathbf{v}\cos\theta_0)^2. \quad (15.4.5)$$

Since the deuteron collisions have a definite axis, the neutron intensity greatly depends upon angle. The cosine of the angle of observation, α, relative to the beam depends upon ψ and is

$$\cos\alpha = \sin\theta_0 \sin\varphi \cos\psi + \cos\theta_0 \cos\varphi. \quad (15.4.6)$$

Except at $\varphi = 0$ and π, the energy is spread by different values of ψ. Combining Eqs. (15.4.1), (15.4.2), and (15.4.6) and noting that φ in Eq. (15.4.2) is to be replaced by α, the maximum neutron energy used in Table I is expressed approximately by

$$E_{n(max)} = \tfrac{1}{2}Mu^2[1 - \rho^2 \sin^2(\theta_0 - \varphi) + 2\rho \cos(\theta_0 - \varphi)]. \quad (15.4.7)$$

Values of $E_{n(max)}$ are tabulated for several values of φ and θ_0 in columns 4, 5, and 6 of Table I, normalized at $\varphi = 0$. Although little can be said of the absolute values, it appears that the angular distribution of neutron energies disagrees with the observed results, for example at $\varphi = 45°$ and $\theta_0 = 45°$. In fact, the neutron energy should peak at the observation angle $\varphi = \theta_0$.

Using Lehner and Pohl's formula[42] for the angular distribution in the c.m. system, the total intensity as a function of φ and θ_0 is obtained by an integral equation as

$$I(\varphi, \theta_0) = 1 + \lambda(\cos^2\theta_0 \cos^2\varphi + \tfrac{1}{2}\sin^2\varphi)$$
$$+ 2\rho(1-\lambda)\cos\theta_0\cos\varphi + 4\rho\lambda\cos\theta_0\cos\varphi(\cos^2\theta_0\cos^2\varphi$$
$$+ 3/2\sin^2\theta\sin^2\varphi) + \rho^2\{[\lambda - \tfrac{1}{2} + (3 - 15\lambda/2)$$
$$\times (\cos^2\theta_0\cos^2\varphi + \tfrac{1}{2}\sin^2\theta_0\sin^2\varphi) + (15/2)\lambda(\tfrac{3}{8}\sin^4\theta_0$$
$$\times \sin^4\varphi + 3\sin^2\theta_0\sin^2\varphi\cos^2\theta_0\cos^2\varphi + \cos^4\theta_0\cos^4\varphi)]\}.$$

(15.4.8)

The ratio of $I(0, \theta_0)/I(\varphi, \theta_0)$ is shown (solid curves) in Fig. 16 for $\theta_0 = 0°$, 30°, and 45°. In no case does the intensity ratio fit the experimental data.

Since the experimental intensity distribution agrees more closely with the moving thermal plasma model, we attempt to show that the neutron energies for $kT = 0$ in column 3 of Table I can be brought closer to the experimental values by assuming that $kT \neq 0$. Taking the observed intensity ratio, $I(0)/(\pi/2) = 1.119 \pm 0.03$, ρ is calculated from Eq. (15.4.4) as $\rho = 0.057$. Also taking the maximum neutron energy in the axial direction as $E_n(0) = 2.89$ MeV, one then uses Eq. (15.4.1), modified to include an additional neutron velocity in the c.m. which is associated with thermal motion of the plasma. The modification of Eq. (15.4.1) leads to $E_n(0) = \tfrac{1}{2}M(u)^2(1 + \rho + x)^2$, where $\rho \equiv v/u$ still represents the c.m. motion and $x = c/u$ is the additional thermal motion term. For $E_n(0) = 2.89$ MeV, $\rho = 0.057$, and $\tfrac{1}{2}Mu^2 = \tfrac{3}{4}Q$, x is calculated to be 0.029. The conditions to satisfy with these parameters are the observed neutron energies and intensity ratios at other angles φ. From the general energy relation $E_n = \tfrac{1}{2}M(\mathbf{u} + \mathbf{v} + \mathbf{c})^2$, one obtains the values listed in column 7 of Table I for $\varphi = \pi/4, \pi/2$, and π radians. These values fall within the shot-to-shot scatter of the data. Similarly, the neutron intensity ratios, Eq. (15.4.4), show that $I(0)/I(\pi/4) = 1.031$ and $I(0)/I\pi) = 1.25$ are within the experimental accuracy limits as shown in column 8 of Table I.

The quantity $x = c/u$, where $u = 2.165 \times 10^9$ cm/sec, gives a value for c of 6.29×10^7 cm/sec. If c is added to u in the c.m. system, the energy spread of the neutron line (2.45 MeV) due to c is $\Delta E_n \simeq 0.13$ MeV. To associate this spread with kT requires some knowledge of the broadening of the 2.45-MeV line by thermal effects. Figure 18 shows Lehner and Pohl's theoretical distribution of the number of neutrons per unit energy interval as a function of neutron energy for a three-dimensional Maxwellian ion distribution at 5- and 10-keV deuteron temperatures. The width at half-intensity is ΔE_n (keV) $= 82.5(kT)^{1/2}$ (keV). Since the time-of-flight

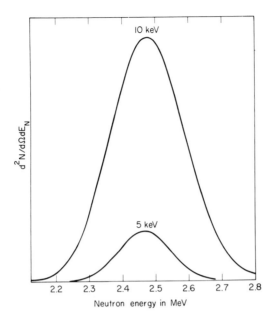

Fig. 18. A theoretical calculation of the number of D–D neutron per unit energy interval for a three-dimensional Maxwellian ion distribution at temperatures of 5 and 10 keV. The 2.45-MeV neutron line is broadened by $\Delta E_n = 82.5(kT)^{1/2}$ (keV), and the peak is shifted to higher neutron energies.

detection apparatus has an intensity threshold, it is clear that the fastest neutrons will correspond to some point along the distribution. There is some ambiguity in associating the observed ΔE_n with a temperature. However, if the half-power point is selected, then the observed $\Delta E_n = 2 \times 0.13 = 0.26$ MeV corresponds to a $kT \simeq 10$ keV. This value is, at best, optimistic because of the detector threshold.

Pulsed neutron measurements show that the neutron pulse width increases with distance from the source. These measurements compare a close-in crystal photomultiplier detector pulse with one placed approximately 15 meters along the axis in a 0° direction. Since neutron production is both time and energy dependent, i.e., $f(E, t)$, it is basically difficult to separate these dependences in any case. However, by combining the squares of the rise times of the two neutron pulses, one obtains a root-mean-square value which, with the flight distance, yields a value for the velocity spread in the neutron source. This velocity spread (ΔV) is hypothesized as reflecting the ion-ion collisional effect in the dense plasma. Values of ΔV for many such data range between 0.8 and 1.0 × 10⁸ cm/sec[1] and would correspond to something like a 24- to 40-keV deuteron interaction in the high-temperature plasma. A three-dimensional Maxwellian ion distribution at temperatures of 4.5 and 7.5 keV could account for the neutron velocity spreads of 0.8 and 1.0×10^8 cm/sec[1], respectively, since the deuteron interactions (equivalent to an energy of ~ 8 times T_i) in the Maxwellian tail produce most of the neutron yield. The lack of a precise detector intensity threshold causes uncertainty in the interpretation; however, many such data suggest that the early ion distribution is strongly peaked in energy and, in time, relaxes toward a thermal distribution. If the ion-ion collisional hypothesis is true, it is the first time such effects have been seen from a hot plasma. An experiment will be performed in the near future to measure the neutron energy spectrum on a single shot.

Neutron measurements reported by Meskan et al.[10] and Maisonnier et al.[12] suggest an accelerated mechanism for neutron production. These geometries, although similar to Figs. 1 and 2, are operated differently.

In earlier low-energy experiments[6] (18-kJ, 20-kV), the acceleration assumption was examined by inserting ZrD and ZrT targets separately into the plasma discharge. The ZrT target, placed on axis ~ 0.25 cm from the expected plasma position, increased the neutron yield by $\sim 50\%$. A second measurement at approximately the same position showed a yield increase about three times. Further measurement was impossible because the target fractured. These increases are actually insignificant since one might expect a factor of 100 on the basis of the relative cross section for D–D and D–T at 10 keV. The largest increase is most probably due to the tritium released from the target during the first discharge. With the ZrD target, many discharges were obtained at various positions near the plasma focus. Better target construction prevented the loss of target. The results showed no significant change in the neutron yield except when the target was placed directly in the focus region, when the neutron yield was drastically reduced. These data are not conclusive but tend to suggest other neutron processes, such as a collision-dominated thermal plasma, rather than one involving a direct deuteron acceleration.

15.4.2. Deuterium–Tritium Experiment

One of the low-energy focus experiments[6] (18-kJ, 20-kV) was redesigned to operate with a deuterium-tritium gas mixture (D–T) rather than pure deuterium gas. The experiment was operated at a reduced condenser voltage (16.3 kV) to reduce the hazard from mechanical or electrical failure. Approximately 29 cm^3 of D–T at NTP was handled per discharge—this corresponds to 45 Ci of tritium per discharge. The vacuum system was designed with a large volume charcoal trap to allow $\sim 95\%$ recovery of the D–T mixture per shot; an ultimate base vacuum of $\sim 10^{-5}$ Torr after each filling was obtained with an oil-diffusion vacuum system. The plasma focus discharge was initially filled to a static pressure of 2–3 Torr. The purity of the D–T mixture was maintained throughout the experiment at $\sim 49.5\%$ deuterium and $\sim 48.5\%$ tritium; the rest was hydrogen. No high-Z contamination > 0.02 of 1% was present.

The motivation for the D–T experiment was twofold, (1) to determine, in comparison to D–D, whether the D–T neutron production scales according to the ratio of the average reaction rates for a Maxwellian distribution, and (2) to determine the effect of the larger mass of the tritium ion on the duration of the neutron pulse. If one assumes that the density and temperature of the D–D plasma focus remain representative of the D–T plasma focus at approximately the same initial filling pressure and applied condenser voltage, and if one further assumes that the D–D plasma is characterized by a Maxwellian distribution, then one can express the relative neutron yields of D–T to D–D for the same number density as

$$N_D N_T \langle \sigma v \rangle_{D-T} / \tfrac{1}{2}(2N_D)^2 \langle \sigma v \rangle_{D-D}, \tag{15.4.9}$$

where the average $\langle \sigma v \rangle$ ratio[46] is $\langle \sigma v \rangle_{D-T}/\langle \sigma v \rangle_{D-D} \approx 200$; this ratio is relatively insensitive to the ion temperature in the 2- to 10-keV range; only the D(Dn) ^{3}He reaction contributes to neutrons. For $N_D = N_T$, the factor $N_D N_T/\tfrac{1}{2}(2N_D)^2 = \tfrac{1}{2}$ which gives a neutron yield ratio of ~ 100. Furthermore, it was expected that the plasma duration might be increased with D–T simply on the basis of the finite Larmor orbit stabilization theory since the Larmor radius of the heavier mass ion, T, would be larger.

The experimental results using a D–T and D filling at approximately the same initial pressure and the same condenser voltage can be summarized: (1) the neutron yield increased to 3.3×10^{11} neutrons/burst for a burst time of 0.22 μsec (half-maximum width of the neutron pulse), and (2) the neutron pulse duration increased from ~ 0.12 μsec for D–D to ~ 0.22 μsec for D–T. The average yield of D–D neutrons was $\sim 4 \times 10^9$ neutrons/burst. For the largest D–T neutron yields, the silver-activated Geiger counting system indicated saturation effects due to pileup. The relative detection efficiencies of the neutron counters for 2.45- and 14-MeV neutrons were approximately 1 and $\tfrac{1}{2}$, respectively. By taking into account the possible saturation effects of the absolute neutron counters and a finite turn-on time of ~ 2 sec, one can upgrade the maximum neutron yield to $\sim 4 \times 10^{11}$ neutrons/discharge, which results in a neutron yield ratio D–T/D–D of about 80 to 100. In a similar experiment, Patou et al.[15]

---

[46] J. L. Tuck, *Conf. Contr. Thermonucl. Reactions, Gatlinburg, Tennessee, 1956*. USAEC Rep. TID-7520, 22 (1956).

obtained a D–T/D–D neutron yield ratio of 30. Furthermore, they record a narrower neutron pulse width for D–T than that with deuterium.

In addition to the above detectors, several other counting methods were employed. An ionization chamber located ~13.7 meters from the discharge gave a very consistent representation of the total neutron yield; its operation depends on the ionization produced by the neutrons in the chamber. The ionization counter was calibrated on the basis of an accepted neutron intensity to produce 1 rem of radiation due to 14-MeV neutrons (1.44×10^7 neutrons/cm^2 = 1 rem). With this calibration, the neutron production was estimated at 8×10^{11} neutrons/discharge. Neutron activation measurements in copper (^{64}Cu) were made ~10 cm from the D–T plasma focus. These results indicated a neutron production of ~2.5×10^{11} neutrons.

15.4.3. Neutron Scaling

The scaling law for D–D neutron production for a given electrical circuit can be related to the initial stored energy, W, to the power α, where α is reported[11,12,14,20] to be 1.5–2.5. These values of α, which appear to vary widely, agree well because the operating parameters of the focus accelerator must be optimized in terms of the applied voltage and initial gas pressure as the energy is increased. Furthermore, the final neutron yield depends also on the plasma volume, time, gas purity, etc. These factors are not easily controlled. In reality, the discharge current is the important parameter that affects the final plasma heating and compression, i.e., $I^2 \approx NkT$. There is general agreement that neutron production is best explained by a high-density thermal plasma. On this basis, using the pressure balance pinch relation, $B^2/8\pi = nkT$, and the D–D reaction rate, $\sim n^2 \langle \sigma v \rangle$, it can be shown that the neutron yield should increase at least with the square of the bank energy, W^2. In this analysis it is assumed that the ion temperature is insensitive to bank energy. In reality, $\langle \sigma v \rangle$ is strongly dependent upon the plasma temperature as T_i^4 for T_i in the 2- to 10-keV range. Also small fractions of high-Z contaminants appear to have a beneficial effect on T_i; however, some uncertainty in the Z effect is evident.[15] Maximum neutron (D–D) yields of 10^{11} neutrons have been obtained in the apparatus[19] described here with a condenser bank energy of 87 kJ; Lafferty et al.[14] report ~1.5×10^{11} neutrons for a bank energy of 250 kJ.

Under good conditions, the reproducibility of neutron production from the plasma focus discharge is within a factor of two. Exceptional reproducibility of ~±15% has been shown by Lafferty et al.[14] for a 250-kJ focus system. Experiments have shown that the initial formation of the current sheath or gas breakdown is the most important and least understood phase

15.4.4. X-Ray Emission

One of the diagnostic techniques commonly used to record the X-ray emission from the plasma focus is the X-ray pinhole camera. Although the pinhole camera is not time-resolved, its spatial resolution is unsurpassed by any other X-ray diagnostic technique. Several X-ray pinhole photographs obtained with a focused discharge driven by a 20-kV, 360-μF condenser bank are shown in Fig. 19 (pinhole diameter 0.25 mm) for $\sim 0.2\%$ mixtures of argon in hydrogen gas. Small fractions of argon enhance the film blackening over that of pure hydrogen. It is clearly seen that the focus development is extremely complex. Streamer formations, apart from the main focus, appear to converge and to form additional intense X-ray sources. The plasma column is broken along the axis. These discontinuous features may be associated with hydromagnetic bouncing of the current plasma while the current remains continuous. On the other hand, z-pinch currents are known to be $m = 0$ unstable. The CE surface appears as an intense source of X rays, and the two-dimensional convergence of the flow pattern in X-ray light is seen.

The usefulness of the photographic technique, aside from recording what has happened, is limited by the nonlinearity of film blackening with photon energy and intensity. For example, the blackening or neutral density, η, of No Screen X-ray film[47] varies from $\eta = 1$ for an incident photon flux of $\sim 10^7$ photons/cm^2 to $\eta = 4$ for 10^8 photons/cm^2. A photon energy increase from 5 to 25 keV requires an additional factor of four increase in photon flux to give the same neutral density. These effects make any quantitative assessment of the photographs difficult.

Quantitative intensity information on soft X-ray emission from the plasma focus is also difficult to obtain by the pulsed detector technique because the spectrum is a function of photon energy, E, time of emission, t, and distance along the focused column, z; i.e., $N_\gamma(E, t, z)$. At present, in most focus experiments, the X-ray emission is sampled from only a small section of the plasma at any one time. If high-Z contaminants are present in the pinched column, the electron temperature or photon energy, intensity, and time of emission will probably vary with the z position. Furthermore, the shot-to-shot reproducibility of X-ray emission is generally poor because of the multiple parameters affecting the emission spectrum. Neutron production, unlike X-ray emission, is singularly

---

[47] L. S. Birks, Unpublished work. Naval Res. Lab., 1967.

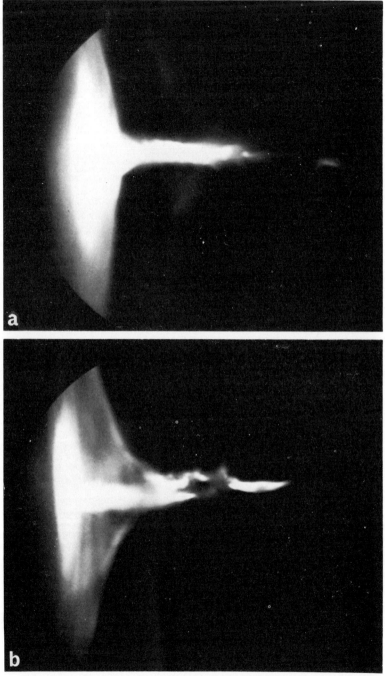

FIG. 19. Typical soft X-ray pinhole photographs of the focus region (pinhole diam, 0.25 mm). The CE is to the left. Condenser energy 67 kJ at 20 kV (1 : 0.75).

15.4. PLASMA DIAGNOSTIC MEASUREMENTS

dependent upon the ion temperature which is directly related to the final current-sheath collapse velocity. No such "simple" thermal process appears to be responsible for the intense X-ray emission.

Two almost distinct sources of X-ray radiation are present in the focus development. The first is the result of electron bombardment of the center positive electrode by the conduction electrons in the current sheath. Part of those electrons can be accelerated to high energies (~ 100 keV) during collapse (see discussion on the voltage generated by the pinch process). The characteristics of this source are similar to those of a high voltage X-ray tube. The second X-ray source is from the focused column. Generally, the focus source is characterized by a soft (low-energy) spectrum which varies with z and t and with the amount of high-Z contamination. The focus is perhaps unstable—this may well account for the inconsistencies in the measured X-ray quantities obtained by different investigators.

There are some indications from other coaxial focus experiments that the X-ray spectrum is dominated by processes[8,10,48,49] other than those sensitive to an electron temperature. For example, Beckner[8,48] has made a detailed examination of the radiation from a deuterium focus near the Cu K_α and K_β lines using a LiF crystal diffractometer. The diffractometer was aligned so as not to include the end of the positive CE. Typical data are shown in Fig. 20. The Cu K_α and K_β line radiations are sufficiently broadened and shifted to indicate that they arise from a variety of ionized species. The absence of helium-like and hydrogen-like line species for an assumed 1-keV plasma temperature was not expected according to a calculation of the time required to populate these ion species. Results similar to those in Fig. 20 were obtained with a negative-polarity CE; i.e., Cu K_α and K_β lines were observed. In either the plus or minus polarity case, it is curious that the copper atoms so readily reach the main body of the focus in a few tenths of microseconds. In a more recent study[49] using a double LiF diffractometer, results similar to those in Fig. 20 were obtained. In addition, when 5% Kr was added to the deuterium filling, the Kr K_α line was observed. If an appreciable number of ions had reached an ionization level as high as Cu XXIV (Kr XXXI), it would have been observed because the calculated K_α line shift for this ionization stage would be significant (>0.2 Å). On this evidence, Beckner[49] concludes that the dominant source of X-ray emission (second X-ray pulse) from the focus discharge is the bombardment of relatively "cold" high-Z atoms by energetic electrons. He further concludes that the absence of

---

[48] E. L. Beckner, *Proc. APS Topical Conf. Pulsed High-Density Plasmas, September 1967*, Los Alamos Sci. Lab. Rep., LA-3770, p. C4–1.

[49] E. L. Beckner, E. J. Clothiaux, and D. R. Smith, Sandia Corp. Rep. SC-DC-68-2155. Sandia Corp., 1968; *Phys. Fluids* **12**, p. 253 (January, 1969).

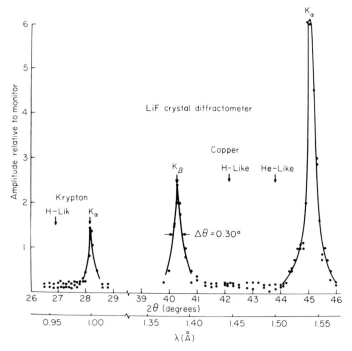

FIG. 20. X-ray emission spectrum in the vicinity of the Cu K_α line from the second X-ray pulse ($t = 2.7$ μsec) of the focus discharge as obtained with a LiF crystal diffractometer. Diffractometer resolution is $0.2°$ at $2\theta = 45°$. (Sandia Lab.)

any helium-like or hydrogen-like spectra is taken as evidence for the absence of a very hot thermal electron plasma.

In another focus experiment (30-kV, 21.6-kJ), Long et al.[9] have shown the absence of any copper species in the main body of the focus. A copper line feature observed between 1.3 and 1.5 Å was located within 5 mm of the positive CE. It is probably due to X radiation rather than to an optical resonance line of hydrogen-like and helium-like copper. No resonance lines in lithium-like copper were observed between 8 and 9 Å.

The evidence of copper contamination in the focus is irrefutable; Beckner's data suggest an ionization of Cu K_α which could be produced only by an electron-atom impact due to high-energy electrons in the Maxwellian tail for electron temperatures of a few keV. The absence of the helium-like and hydrogen-like spectra may be explained as due to an imperfect ionization theory upon which estimates of the time to populate a given ion species is based.[50] Secondary X-ray pulses are common in

---

[50] H. R. Griem, "Plasma Spectroscopy," p. 291. McGraw-Hill, New York, 1964.

15.4. PLASMA DIAGNOSTIC MEASUREMENTS

plasma focus experiments. It is believed that they result from an adiabatic recompression of some part of the focused column, produced by the tube voltage fluctuations. Indeed, joule heating of the electrons in the pinch current may well account for a larger initial electron plasma temperature before adiabatic recompression. The photon spectrum from the secondary X-ray pulses is usually of higher energy than that from the first X-ray pulse.

The difference between Beckner's experiments[48] and those of Long et al.[9] in regard to Cu line radiation is not understood. If line radiation is suggested by Long's[9] experiment within 5 mm of the anodes, then it is possible that the field of view of Beckner's diffractometer may also have included this region. If not, then it is more perplexing how the copper atoms are transported ~ 1 cm into the focus in, say, a tenth of a microsecond (equivalent kinetic energy of 80 keV). Accelerating voltages of this magnitude certainly exist in the focus development, but it seems that the CE polarity change would have affected the transport of copper atoms or ions into the focus region.

The detection of the Kr K_α line in the focus suggests that the krypton atom is already in the body of the focus and does not come from the electrode. Copper atoms might be assumed to be transported similarly by the current sheath to the focused region.

15.4.5. Electron Temperature

One of the principal radiations from a high-temperature plasma is the continuum emission due to electron bremsstrahlung in the ion field. The spectrum, especially at high temperatures, extends deep into the soft X-ray region. The power radiated[40] per unit wavelength per unit volume in the wavelength interval $d\lambda$ is

$$dP_\lambda = 6.01 \times 10^{-30} g \cdot n_e \Sigma (n_i Z^2) T_e^{-1/2} \lambda^{-2} \exp(-12.40/\lambda T_e) \, d\lambda$$
$$\text{watts/cm}^3/\text{Å}, \quad (15.4.10)$$

where T_e is in kiloelectron volts, g is the Gaunt factor, and λ is in angstroms.

From Eq. (15.4.10) we see that the emission decreases rapidly (exponentially) toward the shorter wavelength and with higher electron temperature, and, thus, the spectral intensity over a limited range is a sensitive function of T_e. A measurement of T_e usually consists of comparing the X-ray intensity over different parts of the spectrum.

Absorption filters (filtration method) can be used successfully, and filters with absorption edges can generally give further clarity in investigating certain energy bands. This technique[51] for measuring the electron

---

[51] T. F. Stratton, "Plasma Diagnostic Techniques," Chapter 8. Academic Press, New York, 1965.

temperature is now well established, but certain precautions must be taken to avoid misinterpretation of the results.

One of the greatest problems is to avoid line radiation, especially troublesome in high-powered transient discharges, such as the dense plasma focus where one is likely to find some trace of incompletely stripped high-Z material from the electrodes. To avoid gross errors in the interpretation of the filtered spectrum, a selection of filter material, such as aluminum and copper, can be used. These two materials, for example, with K_α edges of ~ 1.5 keV for aluminum and 8.9 keV for copper will help in sorting or rejecting the K-edge radiations from the plasma. It is highly desirable to make several filtration curves with different materials to obtain good consistency.

In the work of Long et al.,[9] the electron temperature was estimated by measuring the relative intensity of the X-ray radiation through thin foils of nickel, aluminum, and beryllium and comparing the results with that expected from a computed energy spectrum. A p–n junction surface-barrier detector with a depletion layer of $\sim 80\,\mu$ was used. With the addition of 3% argon to deuterium, the electron temperature, T_e, was found to be 1.5 keV. Absolute X-ray intensity measurements indicated an electron density of $n_e > 10^{19}/\text{cm}^3$. In pure deuterium ($P \approx 3$ to 4 Torr), $T_e < 3$ keV.

A similar temperature measurement[2] was made by the author in 1965 with a two-channel X-ray spectrometer using balanced photomultipliers and organic crystal detectors. T_e was estimated as 1–3 keV for pure deuterium using different pairs of absorber materials. Even though the contribution of line radiation from high-Z impurities was assumed small (estimated from different absorber material pairs), the relative transmission results could have been interpreted in terms of a continuum of $Z > 1$.

In a higher energy focus experiment,[52] the electron temperature was determined by measuring the X-ray intensity as a function of absorber thickness from a 1-cm length of the focused column which did not include the CE face. Two 1.5-meter long collimating tubes were employed with fast, lithium-diffused, silicon X-ray detectors.[53] A series of aluminum absorbers up to 116-mg/cm² thick were used in one channel. The detector outputs were time-integrated to obtain the total charge. The energy dependence of the silicon detector was calculated according to

$$Q = (eE/3.5)[1 - \exp(-\mu_{\text{si}}(E)\chi_{\text{si}})] \quad \text{coulombs,} \qquad (15.4.11)$$

[52] P. J. Bottoms, J. P. Carpenter, J. W. Mather, K. D. Ware, and A. H. Williams, Bull. Amer. Phys. Soc. 13 [2], 1543 (1968).

[53] Solid State Radiations, Inc., Los Angeles, California.

where e is the electronic charge, E is the energy of the photon in electron volts, 3.5 eV is the energy required to create a charged pair, $\mu(E)$ is the mass absorption coefficient of silicon, and χ is the absorption thickness (gm/cm^2) of the silicon detector (250-μm depletion depth). The experimental points of X-ray intensity (integrated) versus absorber thickness is shown in Fig. 21 for two capacitor bank energies, 17 and 27 kJ. Comparison of the experimental absorption data and a calculated deuterium bremmsstrahlung spectrum shows a close fit of the 17- and 27-kJ data for electron temperatures of $\sim$1 to 1.5 and 2 keV, respectively. Corrections for the absorption in the vacuum windows (5-mil Be) and the silicon detector response were included in the final integral equation as

$$Q = (e/3.5) \int_0^E N(E')E' \exp[-\mu_1(E')\chi_1] \exp[-\mu_2(E')\chi_2]$$
$$\times (1 - \exp[-\mu_3(E')\chi_3])\, dE' \quad \text{coulombs}, \tag{15.4.12}$$

where $N(E')$ is the bremsstrahlung spectrum for $Z = 1$, and subscripts 1, 2, and 3 refer to beryllium, aluminum, and silicon, respectively. The agreement with the theoretical calculation indicates a negligible contribution from line radiation. Similar results have been obtained with copper absorbers. The deuterium pressure for the 27-kJ, 20-kV discharge was $\sim$7 Torr. If one assumes that the results are due to soft X-ray emission from a pure deuterium plasma 1-cm long and 1 mm in diameter (pinhole camera measurement) which lasts for $\sim$150 nsec, comparison of the theoretical bremsstrahlung emission at the source to the charge collected by the silicon detector for a 2-keV electron temperature suggests a plasma electron density of 4×10^{19}/cm^3.

In principle, it is possible to extract the unknown energy spectrum $N(E)$ from Eq. (15.4.12) knowing the experimental value $Q(\chi)$. The procedure in determining $N(E)$ is to first select a function $N(E)_1$ and then to calculate $Q(\chi)_1$. By comparing $Q(\chi)_1$ with $Q(\chi)$, corrections can be made to $N(E)_1$ until a close fit is obtained. This iteration process is tedious and time consuming, but a spectrum can be obtained. This procedure has been carried out with the experimental data shown in Fig. 21. After many iterations, a spectrum, $N(E)$, has been obtained which agrees with the experimental value $Q(\chi)$ to within $\pm 12\%$. The derived spectrum is shown in Fig. 22. There is some question as to the uniqueness of $N(E)$ obtained in this way and also as to the contribution of high-Z line radiation. However, comparison of the derived spectrum, $N(E)$, with a computed bremsstrahlung spectrum, $N(E)_B$, implies a >1.2-keV electron temperature; the main body of $N(E)$ is shifted to higher energies than is the bremsstrahlung spectrum at $T_e = 1.2$ keV.

The emission spectrum in the X-ray region during a single discharge has been investigated by Long et al.,[9] using a diffraction grating. This

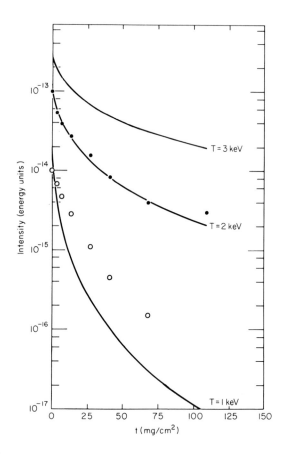

FIG. 21. X-ray intensity as a function of aluminum ($Z = 13$) absorber. The calculated results for a Bremsstrahlung spectrum ($Z = 1$) are shown solid for several assumed electron temperatures. The experimental points are shown for (●) 20-kV (27-kJ) and (○) 16 kV (17-kJ) data.

particular grating was limited to grazing angles of $\sim 2°$ and a maximum wavelength λ of ~ 20 Å. A 2-meter spectrometer was used for wavelengths down to 7 Å, and for $\lambda < 7$ Å, an NPL plane diffraction grating was employed. When 1% of argon or xenon was added to the D_2 gas, a continuum that increased in intensity toward the cutoff wavelength $\lambda_c \approx 7$ Å was observed on the 2-meter instrument. Since T_e (keV) = $6.19/\lambda$ (Å), this sets a lower limit to T_e of ~ 0.88 keV. The addition of larger percentages of impurities produced predominantly free-bound and line radiation, and the plasma was radiation cooled in this case.

For the plasma focus condition,[9] the solution describing the relative

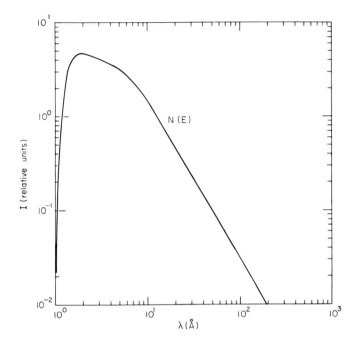

FIG. 22. The derived X-ray energy spectrum, $N(E)$, obtained from the experimental $Q(\chi)$ by the iteration method. Bank voltage = 16 kV (17 kJ).

ion population for a deuterium plasma doped with a few percent of argon indicates that steady state is reached in 0.1 μsec. The calculated steady spectrum is used to explain some of the features recorded with the NPL plane grating at $\lambda < 7$ Å. With a 1% argon-doped deuterium plasma, a strong line feature at $\lambda = 3.9$ Å is seen in four orders. This is accounted for by a blend of A XVIII Ly α at 3.75 Å and A XVII $1s2p'P_1 - 1s^{2}{'S_0}$ transition at $\lambda = 3.96$ Å. X-radiation at 4.2 Å in neutral argon cannot account for this feature. The broad weaker features at shorter wavelengths are interpreted in terms of free-bound continua and higher excited levels of hydrogen-like and helium-like argon.

The general conclusion from the work of Long et al. is that the optical transitions indicate an approach to steady-state excitation in the plasma focus at $n_e > 10^{19}/\text{cm}^3$. There is time for an electron-ion relaxation and a thermal distribution of plasma particles. Furthermore, there is little evidence for copper atoms in the focus region except within 5 mm of the electrode face.

15.4.6. Electron Density Measurement

The Schlieren method has been used by several investigators to estimate the electron density[6,54] and density gradients[55] in the plasma focus. For the electron density, an expression is derived for the mean angular divergence $\langle \theta \rangle$ of a parallel light beam in terms of the refractive index, its gradient, and the boundary-shape factor.

The approximate expression for the refractive index of a free electron gas is given as $n \simeq 1 - \omega_p^2/\omega^2$ where the plasma frequency $\omega_p < \omega$ (light frequency) and the collision frequency[56] $\nu \ll \omega$. This expression for the refractive index is reduced to $n \simeq 1 - (r_e/2\pi)\lambda N_e$, where r_e is the classical electron radius ($r_e = 2.82 \times 10^{-13}$ cm) and N_e is the electron density. Inserting $\lambda = 6.943 \times 10^{-5}$ cm (ruby wavelength), one obtains for the refractive index

$$n \simeq 1 - 2.16 \times 10^{-22} N_e. \tag{15.4.13}$$

The gradient of the refractivity, $\Delta n/\Delta x$, and the plasma boundary-shape factor, $\Delta z/\Delta x$, lead to bending of a parallel light ray. For most plasmas of interest, $n < 1$; hence, the light ray bends toward regions of lower electron density or larger refractive index. The light-ray path in the plasma is actually curved, with a radius of curvature given by $1/R = (1/n)(dn/dx)$. However, in the following derivation of the mean angular deviation, $\langle \theta \rangle$, we assume that the light path is a straight line and that the refractivity gradient, $\Delta n/\Delta x$, averaged over the light path, l, is just $\Delta n/\Delta x \cdot l$. We also assume that the approximation $\theta_{\text{refractive}} \approx \theta_{\text{incident}}$ is valid since the angular deviation is of the order of milliradians.

A plane optical model, shown in Fig. 23, is chosen to calculate the mean angular deviation of a parallel light beam. A refractivity gradient, $\Delta n/\Delta x$, is assumed in the positive x direction, and a boundary shape factor, $\Delta z/\Delta x$, is imposed. Taking the difference, Δt, between light rays 1 and 2 at the plasma exit and letting the angular deviation be approximated by $\theta_D \approx \Delta \rho/\Delta x = (c/n)(\Delta t/\Delta x)$, one obtains

$$\theta_D \approx -\left(\frac{l - 2\Delta z}{n}\right)\frac{\Delta n}{\Delta x} + \frac{2(1-n)}{n}\frac{\Delta z}{\Delta x}, \tag{15.4.14}$$

---

[54] J. W. Mather, Los Alamos Sci. Lab. Rep. LA-3334-MS. Los Alamos Sci. Lab., Los Alamos, New Mexico, 1965.

[55] J. P. Baconnet, G. Cesari, A. Coudeville, and J. P. Watteau, *Bull. Amer. Phys. Soc.* [2], **13**, 1543 (1968).

[56] H. Hora, "Absorption Coefficients of Ruby Laser Radiation in Fully-Ionized Light Elements," 2nd printing. Inst. Plasma Phys., G.m.b.H., Garching, Germany, 1927, 1964.

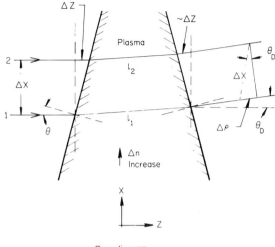

Ray diagram

FIG. 23. The optical model used to calculate the mean angular deviation, $\langle\theta\rangle$, of a parallel light beam; $l_2 = l_1 - 2\Delta Z$; $n_2 = n_1 + \Delta n$.

where the time difference $\Delta t = 1/c(l_1 n_1 - l_2 c_2)$ and c is the velocity of light in vacuum. The valuation of Eq. (15.4.14) depends on the specific shape of the plasma surface.

Examination of the dense plasma focus column in the light of soft X rays leads one to assume that the interesting part of the pinched discharge is rod-shaped, extending a few millimeters along the tube axis, and circular in cross section. It is believed that such a plasma forms at the apex of the converging current sheath.

If a rod-shaped plasma is assumed rather than the model in Fig. 23, it becomes necessary to weight θ_D of Eq. (15.4.14) by $\cos \theta \, d\theta$. For a circular plasma with a boundary of radius a, the following relations can be written:

(1) $l = 2a \cos \theta$, where a is the plasma radius,
(2) $\Delta z = a(1 - \cos \theta)$,
(3) $\Delta z/\Delta x = \tan \theta$, and
(4) $\Delta n/\Delta x = -K \, \Delta N_e/\Delta x$, where $K = r_e/2\lambda^2$.

For a linear decrease in electron density, N_e, over the radius a, i.e., $N_e = N_{e0}(-x/\alpha a + 1)$ where α defines the degree of electron density fall-off, $\Delta N_e/\Delta x$ has been evaluated. Thus, (4) becomes $\Delta n/\Delta x = KN_{e0}/\alpha a$. Making the substitutions (1) through (4) into Eq. (15.4.14) and performing the integration weighted by $\cos\theta \, d\theta$, one obtains for a rod-shaped plasma of circular cross section, for $\alpha = 1$,

$$\langle \theta_D \rangle = -2.14 \times \pi \times 10^{-22} N_{e0}, \qquad (15.4.15)$$

where N_{e0} is the central electron density at $x = 0$.

The optical arrangement for the Schlieren method is shown in Fig. 24.

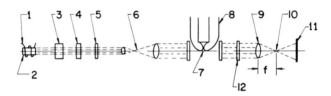

Fig. 24. Optical arrangement for the Schlieren measurement. (1) Pumping lamp; (2) ruby rod (coated one end); (3) Rochon prism; (4) Kerr cell; (5) mirror (reflectance 98%); (6) expanding telescope; (7) dense plasma; (8) discharge apparatus; (9) field lens; (10) pin hole (variable); (11) film plane; (12) ruby filter (narrow band pass 6943 Å).

A ruby laser Q-switched by a Kerr cell gives an ~ 0.1-μsec pulse width at 0.1 MW. Since it is necessary to limit angular divergence of the laser beam to fractions of a milliradian, an expanding telescope is employed; the net angular divergence, θ, of the laser beam at the exit of the telescope is $\theta \approx \Omega_{\text{ruby}} \cdot f_1/f_2$, where f_1 and f_2 are the focal lengths of the eyepiece and objective of the telescope ($f_1/f_2 = 1/15$ where Ω is the natural beam divergence of the ruby). The laser beam passes through a pair of end windows on the discharge tube and is refocused by a 15-cm focal length field lens. Instead of the usual knife-edge Schlieren technique, a pinhole is employed at the focal point of the field lens. The image is then photographed on Polaroid film at 20 cm.

Experimental determination of maximum electron density by the Schlieren method involves finding the largest pinhole aperture that yields a Schlieren effect. One must focus on the interesting part of the dense plasma discharge and record the limiting pinhole aperture, d_1, at which the image vanishes. For a pinhole aperture $d < d_1$ there will be some refracted light rays which miss the pinhole; hence, the image will be dark. For $d > d_1$, all light rays will enter the pinhole and give a uniformly illuminated field. This technique is difficult because the quality of the Schlieren image depends on the contrast of the recording film and the uniformity of the laser beam across the face of the ruby rod. The nonreproducibility of the laser intensity and light pattern adds to the complexity of the image interpretation.

In spite of these problems, it has been possible to determine the electron

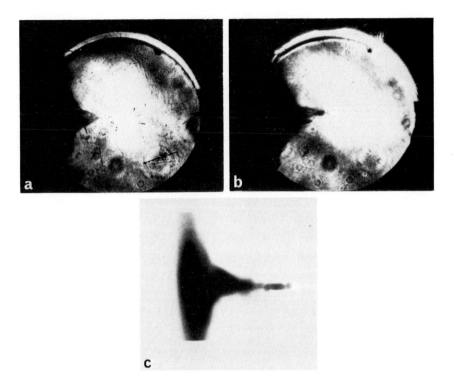

FIG. 25. Schlieren images of the focus discharge for pinhole diameters (a) 1.5 mm and (b) 2.0 mm. A soft X-ray pinhole photograph is shown in (c) (1 : 0.9).

density of the plasma focus. Figures 25a and b show two Schlieren images of the dense plasma discharge (taken during the time of the neutron pulse) for pinhole apertures of 1.5 and 2 mm, respectively. In each photograph the face of the positive CE is at the left. The discharge appears to converge toward the axis in an almost hyperbolic (two-dimensional) fashion. The apex of the converging current sheath usually terminates a few centimeters beyond the CE face, in some instances closer, in others further away. The most interesting part of the plasma discharge is near the apex of this formation. This formation is shown in more detail by a soft X-ray pinhole-camera image in Fig. 25c, taken through a 25-mil-diam pinhole and a thin aluminum absorber (3.14 mg/cm^2). It is that region, at the apex of the converging current sheath, which gives the evidence to support the assumption for the rod-shaped plasma. The images indicate that the interesting part of the discharge is still present for pinhole diameters of 1.5 and 2 mm. The 2-mm diam would correspond to an experimental mean angle of deviation of 6.7 mrad; i.e., $\langle \theta_D \rangle_{exp} = d/2f$. Solving for N_{e0} according to Eq. (15.4.15), one obtains $N_{e0} = 10^{19}/\text{cm}^3$ for the electron density. This represents a minimum electron density; a maximum value has not been obtained.

15.4.7. Spatial Stability of the Focus

The influence of small dc "axial" magnetic fields upon the stability[23] of the focused plasma column is shown by soft X rays. The material damage to the CE and the intense X-ray emission at its face as a result of electron bombardment is considerably reduced. Although the final heating of the ions and electrons is inhibited, as expected, due to a trapped magnetic field, the spatial stability of the focused column offers a greater possibility for scaling to higher energies.

Figure 26 is a schematic representation of the 67-kJ focus and the internal magnetic field arrangement. The shape of the magnetic field,

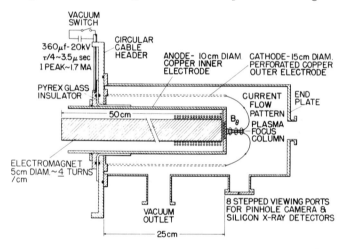

FIG. 26. The 67-kJ plasma focus and the stabilizing electromagnet arrangement.

exhibited by iron filings, is shown in Fig. 27. The numbers indicate approximately the dipole magnetic field magnitudes at various positions downstream from the CE and on the CE surface in the annular region of the plasma accelerator. The axial component of the magnetic field decreases from ~120 gauss at the CE face to ~42 gauss at an axial position 3.5 cm from the face for a 14-ampere dc magnetizing current. Currents of up to 84 amperes produce magnetic fields of ~720 gauss at the CE face.

It is clear from Fig. 27 that the magnetic field varies in r and z. The radial magnetic field between the coaxial electrodes in the accelerator section seems to perturb the dynamics of the axisymmetric current sheath

15.4. PLASMA DIAGNOSTIC MEASUREMENTS

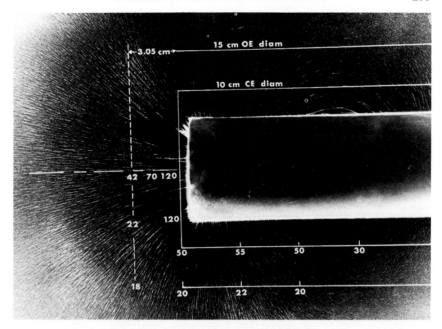

FIG. 27. Stabilizing magnetic field contours shown by iron filings. The numbers refer to the field magnitude for a 14-ampere dc current. The solid lines represent the boundaries of the inner and outer electrodes.

only slightly during acceleration. One can imagine the radial magnetic field lines that cross the annulus frozen to the inner and outer copper electrodes at their respective locations while the field lines in the annular space are distorted and swept along by the advance of the axisymmetric current piston. In this way all the field lines encountered by the current sheath become trapped in the focus column. Some "cushioning effect" or magnetic insulation is provided by the field lines that are compressed to the surfaces of the inner and outer electrodes.

The magnitudes of the compressed magnetic field obtained by axial probes 3.15 and 4.5 cm from the electrode face are 150 and 100 kG, respectively, for an applied field, B_0 of 720 gauss and a discharge voltage, V_0, of 20 kV. For $V_0 = 15$ kV and $B_0 = 360$ gauss, the compressed field at 2.3 cm is 125 kG.

The soft X-ray pinhole photograph, Fig. 28, demonstrates the influence of various magnetic field strengths upon the plasma focus stability of a hydrogen–argon (~0.2% argon) mixture. Figure 28a is the normal plasma focus; i.e., focus with no magnetic field. The successive photographs, (b) through (f), are for axial magnetic fields of 120, 240, 360, 480, and 720 gauss, respectively. Accompanying each X-ray pinhole photo-

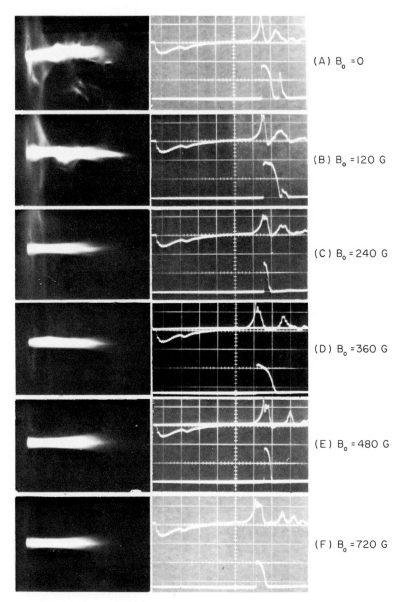

FIG. 28. Soft X-ray photographs (0.25-mm pinhole diam) and the corresponding dI/dt and soft X-ray pulse oscillograms as a function of the stabilizing magnetic fields for the 67-kJ, 20-kV plasma focus. Note the diminution in X-ray intensity at the CE face (left in the figure) as B increases.

graph is the voltage waveform (upper trace) and the response of a lithium-diffused silicon X-ray detector (lower trace) on an expanded time base (sweep speed 0.5 μsec/division). The silicon detector views a 1-cm-diam circular portion of the plasma focus column ~1 cm from the CE face.

In Fig. 28a ($B = 0$), one not only observes the plasma column region, but also regions of plasma fluctuation near the CE face which appear to converge in or near the main column. At 3–4 cm from the electrode, a region of violent mixing of plasma streams is evident. These plasma fluctuations and streamers usually generate intense bursts of X rays of greater hardness than those from the plasma column. The shot-to-shot reproducibility of such phenomena is poor. In fact, it was these observations that led to the idea of stability control utilizing magnetic fields. With the addition of a magnetic field (Figs. 28b–f), enhanced stability of the plasma focus column is achieved. For low field values, photographs (b) and (c), some plasma streamers do remain, but by and large, compared to (a), a great improvement is effected by the magnetic field.

Another effect accompanying the "normal" formation of the plasma focus is the cavitation and erosion of the CE caused by the high electron current bombardment and the momentum recoil of the hot dense plasma against the electrode. The erosion is severe at high energies and has been controlled somewhat by inserting a tungsten disk in the center of the copper electrode; however, erosion, and cavitation continue. At the same time, an intense, high-energy X-ray emission is produced at the electrode surface. With the application of the magnetic field, the X-ray emission is reduced by several orders of magnitude and the cavitation proceeds at a small controlled rate.

The decreased surface destruction with the field can be explained by supposing that the magnetic lines, frozen into the metal surface at their respective locations, are bent and compressed against the electrode face by the collapsing current sheath to form a magnetic cushion similar to that described previously. During the convergence of the plasma current sheath on axis, the field line shape resembles a "funneling" of lines from the electrode surface into the focused column. This shape not only tends to spread out the impact of electrons on the electrode surface but appears to substantially reduce surface erosion and X-ray emission. Nevertheless, with a magnetic field the impact from the plasma recoil is still evident along the axis though much less serious. Plasma pressures (nkT) of the order of 10^5 atmospheres are typical in the focused column.

To gain further insight into the effect of magnetic stability, the correlation between the time history of X-ray emission and the interelectrode voltage signal is considered. The event of interest occurs ~3 μsec after breakdown, when the first plasma current compression occurs. This compressional process generates a back voltage, V, due to the rapid change in the circuit inductance, $V = i\, dL/dt$ (see Section 15.3.5). Since $>95\%$

of the tube current is carried in the plasma column, any subsequent time changes in the radius or length of the column also produce fluctuations in the voltage waveform. Note that resistive changes in the plasma column can also produce similar voltage fluctuations to those of $i\,dL/dt$. In the oscillograms of Fig. 28, several voltage fluctuations occur after the main compression pulse. Depending upon the included magnetic field, these fluctuations may, or may not, lead to a secondary X-ray emission. For a zero or small B field, the secondary voltage pulses appear to be correlated with a photon emission which is substantially harder than the photon energy emission at the time of the first compression. This observation leads to the speculation that the secondary X-ray pulses are from a higher temperature plasma produced by further adiabatic recompression. This hypothesis is further supported by the fact that the secondary X-ray pulses are inhibited by increases in the B field while the voltage fluctuations are hardly diminished. With a deuterium plasma and with that magnetic field to stabilize the plasma column, neutron production is reduced an order of magnitude. In this instance, the trapped magnetic field hinders the recompression and substantially reduces the final ion and electron temperatures and particle density. This can be justified qualitatively from the pressure balance relation $B_\theta^2/8\pi = nkT + B^2/8\pi$, where B_θ is the azimuthal self-field of the current, B is the included stabilizing magnetic field, and nkT is the plasma pressure.

15.4.8. Theoretical Study

15.4.8.1. Plasma Lifetime.

The plasma focus lifetime appears not to be dominated by hydromagnetic effects. A lifetime of ~ 1 nsec would be expected based on the plasma radius and local sound speed. Finite orbit theory would be reasonable to account for the lifetime on the basis of the ion gyration radius; however, at best, the B_θ field distribution within the plasma is linear and, in the limiting case of a thin shell, it is localized at the current sheath boundary. Whether it makes sense to talk about finite orbit effects seem questionable. The stabilizing effect of finite Larmor orbit radius has been shown[57] for low-β plasmas, i.e., for plasma for which $p \ll B^2/8\pi$. Extending the theory to high-β plasma is questionable and more so for the case of high-density plasma where collisional effects are important. In spite of uncertainty, a condition for stabilization[58] due to finite orbit effects requires that the Larmor radius a must be of the order

---

[57] K. V. Roberts and J. B. Taylor, *Phys. Rev. Lett.* **8**, 5 (1962).

[58] B. R. Suydam, Los Alamos Sci. Lab. Rep. LA-3260-MS. Los Alamos Sci. Lab., Los Alamos, New Mexico, 1965.

of the plasma radius, especially for the $m = 1$ mode. The condition for stability for the $m = 1$ mode, $k_{\text{wave number}} = 1/r$ is

$$(a/r) > 2(2\alpha)^{1/2}/(1 + \alpha), \qquad (15.4.16)$$

where α is a factor between 1 and 2. For dense plasma focus parameters, $r \sim 2.5 \times 10^{-2}$ cm and $B_\theta \approx 10^6$ gauss, a is much less than r. For the $m = 2$ mode, $k = 2/r$, the condition is

$$(a/r) > (2\alpha)^{1/2}/(1 + \alpha). \qquad (15.4.17)$$

Thus higher m modes are more easily stabilized. In the derivation of Eqs. (15.4.16) and (15.4.17), it was assumed that $n \equiv (1/N)(dN/dr) = 1/r$, i.e., the density peaks at $r = 0$. A more realistic model might deal with a density distribution function which falls sharply at the sheath boundary, i.e., $N = 1/r\delta$, where $\delta \ll 1$. It is clear that the present theory is inadequate to discuss plasma focus stability.

One can calculate the drift velocity of an ion along the pinched column due to a magnetic field gradient according to the relation[36] $v_D/v_\perp = (a/2)(\nabla_\perp B/B)$, where a is the radius of gyration and $\nabla_\perp B$ is the gradient of the scalar B in the plane perpendicular to $\mathbf{B}$; $v_\perp$ is approximately the thermal ion speed. For $\bar{B} = 10^6$ gauss, plasma radius $\sim 2.5 \times 10^{-2}$ cm, and particle energy ~ 5 keV, one obtains $v_D/v_\perp \simeq 0.3$. The drift time, for a plasma length of ~ 1 cm is ~ 50 nsec. This result is in reasonable agreement with experimental observation. Similar results can be obtained by associating a microelectric field[59] in the plasma of the order of $kT_1/e\lambda$, where λ is the distance over which $E_\perp$ has a constant value. The ions will then drift according to $\mathbf{E} \times \mathbf{B}/B^2$. It can be shown that this formulation reduces approximately to the above result. Comisar,[16] on the other hand, has shown that the theoretical lifetime of the plasma focus can be explained by including pinch curvature in the usual sausage instability theory. It then becomes unnecessary to invoke new mechanisms such as finite Larmor radius stabilization or gradient B drifts.

The observed dense plasma focus duration is explicable, perhaps, in terms of the z-pinch with increased radial stability. But how does one explain the duration in terms of the losses in the z direction for a finite length z-pinch? From a simple estimate of particle transit times along the pinch axis, it appears that any flute-type instability wave leaves rapidly before any appreciable radial growth occurs. The average electron velocity in the pinched plasma is estimated to be $\sim 3 \times 10^8$ cm/sec corresponding to ~ 27 eV; it seems probable that a "cold" electron flow through the plasma region may offer some stabilization.[17] An analysis similar to that

[59] T. K. Fowler, Unpublished work. Lawrence Radiation Lab., Livermore, California, 1965.

outlined by Longmire[60] for an infinitely long pinched conductor appears to be warranted. The solution of the Liouville equation, without use of the hydromagnetic or finite orbit effect approximation may give some idea about the plasma containment along the axis of a finite length pinch.

15.4.8.2. Computer Study. A preliminary study[61,62] of the dense plasma focus discharge has been made using a two-dimensional magnetohydrodynamic code. Some results of this study are shown in Figs. 29 and 30 which pertain to the most interesting and least understood region of the dense plasma discharge. The inclusion of this work is not to be construed as a final or complete discussion of the phenomena but rather as showing some of the plasma sheath characteristics that are otherwise difficult to obtain. Much of the information calculated, such as the rate of tube inductance change, sheath velocity, kinetic energy, and time history of the sheath resistance during current sheath passage in the accelerator has been omitted because the experimental results and the calculations agree well. (See Section 15.3.5.)

The two-dimensional formulation utilizes Eulerian hydrodynamics with variable grid mesh size. In the regions of interest, smaller mesh is used. The code treats both ions and electrons as fluids with temperatures T_i and T_e; assumes that the outer electrode is impervious and acts as a heat sink, i.e., that it is a boundary on the problem; assumes that the electrons lose energy by radiation; includes thermal conduction by electrons but not by ions; and assumes that the ohmic heating is due to the ordinary resistivity given by Spitzer's $T_e^{3/2}$ relationship.[36] The discharge is started at the breech by providing a conducting layer at ~ 1 eV temperature. Maxwell's equation determines the magnetic fields. It is assumed that no material comes from the surface of the electrodes to contribute to pressure balance.

The gas sheath profiles show that the current flows mostly in the forward part of the sheath and that very little current is in the gas near the wall. This result has been observed with magnetic probes.[2] The calculation indicates that the sheath thickness is ~ 1 cm midway in the annular region and that the current layer is canted backwards from the inner to the outer electrode. T_e and T_i at the midpoint of the sheath increase from about 18 to 25 eV during the acceleration phase; the sheath velocity is ~ 10 cm/μsec.

Figure 29b shows an r, z plot of the current trajectories at and beyond

[60] C. L. Longmire, "Elementary Plasma Physics," Vol. IX. Wiley (Interscience), New York, 1963.

[61] J. LeBlanc and R. Wilson, Unpublished work. Lawrence Radiation Lab., Livermore, California, 1967.

[62] B. R. Suydam, Unpublished work. Los Alamos Sci. Lab., Los Alamos, New Mexico, 1967.

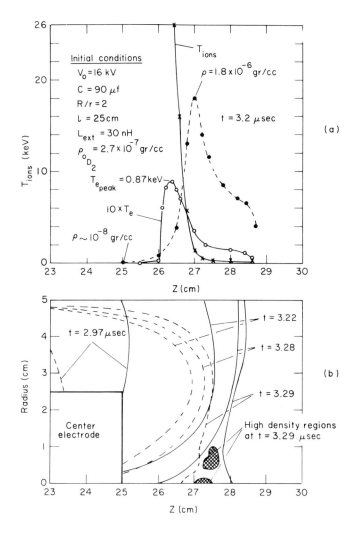

FIG. 29. Results of the 2-D hydromagnetic code computation showing (a) T_i, T_e, and mass density, ρ, versus position z and (b) the current trajectories during the convergence of the current filament toward the axis at four times. The solid and dotted lines correspond to the leading and trailing edges, respectively. The shaded areas represent "on" and "off" axis high density regions.

the end of the CE. The solid and dotted lines refer to the forward and rear portion of the current sheath, respectively. Four such configurations are shown; one prior to collapse at 2.97 μsec and three at 3.22, 3.28, and

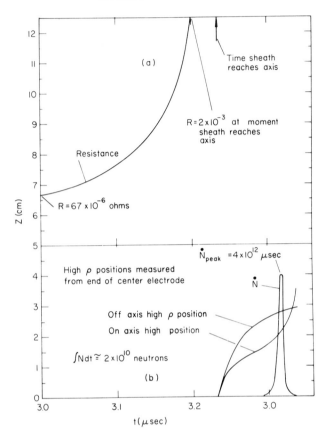

FIG. 30. The position of the high density regions and the neutron production rate, D–T (b) and plasma sheath resistance (a), as a function of time. Note similarity between (a) and Fig 10.

3.29 μsec during collapse. It is curious that two high-density regions form, one on- and one off-axis. No physical explanation is offered.

Figure 29a shows the z-position dependence of the gas density, the electron temperature, and the ion temperature on-axis at 3.2 μsec. The gas density ($\sim 10^{-8}$ gm/cm^3) near the CE face actually falls below ambient (2×10^{-7} gm/cm^3) but increases to a peak of 1.8×10^{-6} gm/cm^3 at approximately 2 cm from the electrode. If the mass density near the electrode is this low, then current carrier starvation would cause large voltages to be generated. It appears experimentally that sufficient gas may be produced by the vaporization of the electrode surface due to the inward sweeping of the collapsed current across the electrode. The peak density of 1.8×10^{-6} gm/cm^3 is low by a factor of ~ 30 if the measured density[7,54] of $\sim 2 \times 10^{19}$ gm/cm^3 is correct. The peak electron temperature of 0.87 keV is low compared to experimental values of about 2 to 3 keV. The plasma focus discharge is a copious emitter of X rays, some of which come from a hot plasma. The two-dimensional code, at least at this stage of development, cannot produce high electron temperatures partly because the collisional mean free path is much longer than the plasma dimensions; i.e., $\omega\tau \gg 1$. At this stage, the assumption of a pure hydrodynamic electron heating process is probably not valid. This result, however, does not detract from the overall success of the code, rather it means that other heating processes[21] must be invoked for the last moments of collapse to achieve high electron temperatures. The ion temperature at 3.2 μsec drops from 26 to ~ 2 keV within $\sim \frac{1}{2}$ cm. Part of the ion curve not included here extends into low-density regions behind the current front to larger temperature values. The computed temperatures in these regions are not believed to actually occur; in the future, inclusion of ion conduction and viscosity terms should reduce T_i behind the sheath considerably. At an axial distance $z \sim 2$ cm, $T_i \sim 1.5$ to 2 keV.

Figure 30a shows the time history of neutron production and the trajectories of the off- and on-axis high ρ regions. The z coordinate is measured from the CE. There is general motion of the dense region; the on-axis region at first moves slower than the off-axis region, but eventually the velocity of the off-axis region decreases almost to zero. The on-axis velocity at the time of neutron emission increases to $\sim 8 \times 10^7$ cm/sec and later becomes larger. This motion could produce the so-called moving "boiler"[5] mechanism of neutron production. In this mechanism, "boiler" refers to a hot thermonuclear plasma, and it is the axial translation of this hot plasma which gives rise to the moving "boiler" concept. The calculated production of 2×10^{10} neutrons agrees well with the experimental results ($\sim 10^{10}$) within a factor of 2 to 3. The neutron production was calculated by integrating $n^2 \langle \sigma v \rangle$ over the axial region extending from the center electrode. The contribution to the neutron pulse is small in the regions of high T_i and low density. By and large, the emission is produced near $z \approx 2$ cm for a $T_i \approx 1.5$ keV and a mass density of $\sim 1.8 \times 10^{-6}$ gm/cm^3.

Several two-dimensional codes now exist with various degrees of sophistication. Besides the work reported above, computer codes at the Kurchatov Institute in Moscow[61,63] and at the Culham Laboratory in England[64] include most of the plasma physics parameters. Many questions concerning the plasma boundary conditions at the electrode surface and plasma heating mechanisms are known experimentally to be important,

---

[63] V. S. Imshennik and V. F. D'yachenko, Unpublished work. Inst. of Appl. Math., Moscow, USSR, 1968.

[64] K. V. Roberts and D. Potter, Unpublished work. Culham Lab., England, 1968.

but the exact formulation of these boundary conditions in the code is difficult.

A different computing technique, PIC (particle in cell), is used by Butler, et al.[41] to study the motion of a high-current discharge in coaxial geometry (negative CE). The numerical model utilizes two-dimensional hydrodynamics modified to include magnetic effects. The PIC method uses a Eulerian mesh of cells with Lagrangian particles to describe the fluid. The flow region is subdivided into rectangular cells, each characterized by average values of such two-dimensional flow variables as pressure, density, internal energy, and velocity. This formulation does not include all the desired plasma physics parameters; it does, nevertheless, describe current sheath motion and demonstrates the plasma flow beyond the terminus of the negative CE. The two-dimensional convergence of the on-axis plasma flow produces a high-density region similar to that observed in plasma focus.

15.5 Summary

The plasma focus discharge has been discussed largely from the experimental point of view. Some of the experimental information can be predicted theoretically; however, the processes involved in breakdown and development of the current sheath, and the heating of ions and electrons during the fast compressional process leading to the focus and during the quasisteady period of the focus are not well understood. The two-fluid, two-dimensional hydromagnetic models are quite successful in predicting some qualitative and quantitative results, but more theoretical effort is needed. With the advent of more sophisticated two-dimensional hydromagnetic codes, better understanding of the focus will result. Inclusion of other plasma heating mechanisms and real boundary conditions at the electrodes is needed for a closer fit to the experimental results. One-dimensional models[63] have been found inadequate to explain Filippov's et al.[4] focus results.

Further experimental measurements are needed to learn more about the collapse mechanism. Schlieren photography, laser scattering, and neutron and X-ray production studies offer the best diagnostics and can reveal important information without disturbing the focus. For example, a coherent laser scattering experiment is expected to yield time-dependent information about the electron and ion temperatures and electron densities along the focused column. With the stabilizing effect of small dc magnetic fields, the scattering experiment would make perhaps the most important contribution to an understanding of the focus. Detailed X-ray and neutron measurements as a function of space and time can also reveal something

15.5. SUMMARY

about the electron and ion velocity distributions. The inclusion of simple magnetic fields does inhibit plasma heating but leads to a more stable focused column. It appears that a quasi-stationary plasma configuration is created; however, there is evidence that the entire plasma column is not produced simultaneously.

The overall interest in plasma focus stems generally from a purely scientific interest in dense hot plasma, $n \geq 10^{19}/\text{cm}^3$ and $T > 1$ keV. Morozov's recent study[35] of steady-state flow reactors (controlled thermonuclear reactors) suggests, however, in the author's opinion, a possible controlled thermonuclear reactor application of the plasma focus principle. At present, the focus can also be used to study short-lived neutron activation of the elements in the subsecond region and the production of highly ionized states of heavy Z ion species. The intense neutron production can also be used to supplement the nuclear fission reactor neutron spectrum for medical research, and, more recently, the application of intense pulsed neutron sources to the assay of fissionable materials in the nuclear safeguards program[65] has seemed possible.

---

[65] G. R. Keepin, Los Alamos Sci. Lab. Rep. LA-3802-MS. Los Alamos Sci. Lab., Los Alamos, New Mexico, 1967.

16. PLASMA PROBLEMS IN ELECTRICAL PROPULSION*

16.1. Introduction

16.1.1. Criteria for Electrical Propulsion

The involvement of plasma physics and technology in the area of space propulsion has occurred principally because of two determining facts: (1) With the sole exception of nuclear explosions, the electrical or electromagnetic acceleration of conducting matter seems to be the only feasible way of bringing rocket exhausts to the very high velocities necessary for deep space flight with payloads of men and their support equipment. (2) Ionized gas, or plasma, is by far the most convenient conducting material to use, since by virtue of its usually low density, it can be brought to the requisite velocities without the expenditure of prohibitive increments of energy.

Several excellent developments of the rationale for electrical propulsion are available.[1,2] We will only briefly summarize the standard argument here.

Suppose that a certain payload for a deep space mission has been brought into low earth orbit by conventional means. A certain additional increment of velocity $\Delta \mathbf{v}$ must now be given this payload in order to send it to, say, some other planet. Initially, the system mass m_0 consists of the useful mission mass m_m plus the mass m_p of propellant necessary to achieve the necessary velocity increment. If $\Delta \mathbf{v}$ is produced through ejection of exhaust at a constant velocity $\mathbf{u}_e$ relative to the vehicle, we can write Newton's law as

$$\mathbf{u}_e \, (dm/dt) + m \, (d\mathbf{v}/dt) = 0 \qquad (16.1.1.\text{a})$$

and integrate it directly to obtain, for the velocity increment after expenditure of all the propellant mass,

$$\Delta \mathbf{v} = \mathbf{u}_e \, \ln(m_0/m_m) \qquad (16.1.1.\text{b})$$

---

[1] E. Stuhlinger, "Ion Propulsion for Space Flight." McGraw-Hill, New York, 1964.
[2] R. G. Jahn, "Physics of Electric Propulsion." McGraw-Hill, New York, 1968.

---

* Part 16 by Ralph H. Lovberg.

or
$$m_m/m_0 = \exp(-\Delta v/u_e). \tag{16.1.1c}$$

We can now see that in order to avoid orbiting excessive amounts of propellant relative to useful payload, one should employ exhaust speeds u_e comparable to or greater than Δv for deep space missions. For example, u_e for the "exotic" chemical propellants is about 5×10^5 cm/sec, while Δv for a mission from Earth orbit to Jupiter and return is about 6×10^6 cm/sec. The necessary chemical propellant mass would be about 10^5 times greater than that of the mission payload!

The electrical acceleration of mass is, of course, a nearly ancient art; furthermore, it is almost trivially easy to obtain velocities between two and three orders of magnitude higher than Δv for the Jupiter mission. One might imagine, then, that the problem of propellant mass could be entirely eliminated by the use of conventional (but high current) particle accelerators as rockets. This is impractical, however, for the simple reason that the rocket power (which now must come from an electrical supply) is equal to the thrust ($\dot{m}u_e$) multiplied once again by u_e. As a result, one discovers that the production of even marginally useful thrust levels from beams having u_e much in excess of 10^7 cm/sec requires power supplies so massive that the advantage of low propellant mass is entirely lost.

An optimization of the exhaust velocity is possible, as one might suspect from the above arguments. It turns out, not surprisingly, that one can perform a given mission (in the sense of taking a given m_m to some assigned target and returning) with a minimum m_0 by selecting that value of u_e which makes the mass of propellant and power supply roughly equal. There is thus a premium on achieving low specific power supply mass (in mass per unit power), as well as high efficiency of all electrical components. The details of an exact optimization involve the nature of the orbits, the length of rocket burn, and other factors; nevertheless, whether the problem is done simply or exactly, the answer for any interesting (or reasonable) space mission, and given the present or near future state of power supply development, is that
$$10^6 < u_e < 10^7 \quad \text{cm/sec}.$$

It is clear that even for the Jupiter mission, which is the most ambitious of those being seriously considered at the present writing, the useful payload fraction will not be too unfavorable if electrical propulsion is employed.

16.1.2. Plasma Parameters in Electrical Propulsion Systems

The requirements upon a plasma suitable for electrical propulsion differ markedly from those for other plasma technologies, e.g., controlled fusion;

16.1. INTRODUCTION

indeed, in comparison to fusion requirements they are nearly opposite. It is necessary for propulsion that the plasma be well enough ionized and hot enough to be a good electrical conductor (in a sense which will be made clear), but also that it not be allowed to carry away into the directed exhaust beam any substantial internal or nondirected energy, such as kinetic heat, excitation energy, or unnecessary ionization energy, since all of these are nonpropulsive losses, or "frozen flow" losses, in rocket parlance. It is, of course, possible to recover the energy contained in the randomly directed velocity components of a streaming plasma by use of a nozzle, as is the practice in conventional rocketry; however, this involves interaction of the hot plasma with a material wall, and will usually result in severe erosion of a nozzle.

Reduction of the fractional energy loss attributable to frozen flow is usually accomplished by the selection of relatively heavy atoms for the plasma. One may recall that system design constraints such as the power supply specific weight and the mission length combine to determine an optimum exhaust speed u_e, and that this speed must apply to whatever medium is ejected; consequently, the kinetic energy per ejected ion will be in direct proportion to the atomic mass. However, ionization potentials do not depend strongly on mass, and thus, the fraction of input energy necessary to ionize becomes small for heavy materials. The following example, calculated for an exhaust speed of 5×10^6 cm/sec, makes the point clear:

Gas	KE (eV)	W_i	Loss (%)
H	12.5	13.6	52
Ar	500	15.6	3
Cs	1700	3.9	0.23

Energy lost through excitation and radiation tends to be of the same order, or less per ion, and so all of these losses can be made tolerable through the use of heavy propellants.

Having considered in a rough way the plasma temperatures and species which are optimal for propulsion, we may also attempt some estimate of the density at which one might work. Consider the thrust of a rocket having a nozzle area A:

$$F = nmu_e^2 A, \qquad (16.1.2)$$

where n is the density of particles of mass m and directed velocity $\mathbf{u}_e$.

(We assume only one species, monoenergetic, for simplicity.) For argon at $u_e = 5 \times 10^6$ cm/sec[1], we have

$$n = 6 \times 10^8 F/A,$$

where F is in dynes and A in square centimeters. The value of F/A which it will be necessary to achieve is not very clear; however, if one uses the usual rule of thumb that the vehicle acceleration must exceed 10^{-3} g and also inserts the likelihood that the vehicle mass will be of the order of 10^4–10^5 kg, a thrust of 10^7–10^8 dyn as appears necessary. A reasonable upper limit to engine area might be 10^5–10^6 cm^2; therefore F/A will lie within an order of magnitude of 100 dyn/cm^2. The corresponding average plasma density in the exhaust is then

$$n \geqslant 10^{11}/\text{cm}^3.$$

This estimate applies, of course, to steady-state accelerators. If the engine is operated impulsively, the above figures must be divided by the "duty cycle," or ratio of impulse time to repetition period. A typical duty cycle for pulsed engine designs is 10^{-4}, giving

$$n \text{ (pulsed)} \simeq 10^{15}/\text{cm}^3.$$

In summary, then, we may conclude that the field of electric propulsion is concerned with plasma of medium to heavy elements, fairly well ionized, and having electron and ion densities in the range of 10^{11}–$10^{15}/\text{cm}^3$.

16.1.3. Categories of Electrical Propulsion Systems

Electrical propulsion engine concepts are quite numerous, and we will not discuss them all here. A classification system in common use divides accelerators into categories which are characterized by the mode of coupling between electrical system and gas.

Thus, *electrothermal* engines are those in which the input power serves only to heat the working gas thermally, by such transfer devices as hot resistors or electric arcs. From this point on, the device is a conventional rocket, with the propellant expanding in a nozzle in order to convert thermal to directed kinetic energy. It is not even necessary, in the case of "resistojets" that the gas be ionized. Since these are not properly plasma devices in the sense which interests us here, we will not treat them further.

Electrostatic engines or *ion engines* accelerate plasma by directly applying electric fields to ions, much as is done in nuclear particle accelerators. Once the ions have been ejected from the region of high electric fields, an admixture of electrons is provided, and the exhaust becomes a neutral plasma. Since the acceleration region must contain ions alone, there are formidable space charge problems if an even nearly adequate thrust density is to be achieved. Notwithstanding, ion engines of high efficiency, long life, and acceptable thrust have been developed.

Electromagnetic engines are those which apply force to the propellant through the interaction of magnetic fields with electric currents flowing in the plasma. Within this category are found many subspecies, which may involve steady-state or impulsive operation, externally applied magnetic fields, induced plasma currents, or currents introduced through electrodes. The reader is again referred to the quite complete summary of the field by Jahn[2] for details of these and other concepts.

16.2. Electromagnetic Propulsion as a Problem in Magnetohydrodynamics (MHD)

16.2.1. Equations and Useful Approximations

In a certain important class of plasma experiments associated with propulsion technology, it has only been necessary (or at least most convenient) to consider the plasma as a single fluid, endowed with the properties of density, electrical conductivity, and pressure. In this case, we are studying magnetohydrodynamics (MHD) rather than plasma physics since the latter is exactly the study of ionized gas as a multicomponent fluid.

Nevertheless, the MHD approach is nearly sufficient for the experimental and even theoretical study of many plasma applications.

The object of any experimental study of an MHD device whose object is propulsion will naturally be the transfer of momentum and energy from the electrical supply system to the working fluid; in particular, one usually desires to know the efficiency with which this transfer is accomplished.

If we approach the problem with the greatest generality, we first employ the set of equations which describe the conservation of mass, momentum, and energy, as well as the equation of state deemed appropriate for the plasma. Maxwell's equations must always be invoked, and also suitable transport coefficients expressing electrical and thermal conductivity, radiation, and viscosity as functions of the parameters of the plasma. It is usually impracticable to employ the most general set of equations in the theoretical analysis of an MHD system, or in the interpretation of experimental data. Generally, however, it is possible to exploit one or more ways in which a system under experimental study approaches extreme limits in certain parameters, making them either negligible or completely dominant. Equivalently, in theory, one may arbitrarily set certain parameters, or their space or time derivatives to zero without removing the problem too far from reality. Such convenient limits will be assumed in most of the work described here.

We may take as an example the equation of momentum transfer,

$$\rho(\partial \mathbf{v}/\partial t + \mathbf{v} \cdot \nabla \mathbf{v}) = \mathbf{j} \times \mathbf{B} - \nabla p + \mathbf{f}_v, \quad (16.2.1)$$

where ρ is the fluid (plasma) mass density, p is the pressure, and $\mathbf{f}_v$ is viscous force. The left-hand parenthesis is the convective velocity derivative. A plasma accelerator designed as a thruster will inevitably involve only one direction of acceleration, and so only one component of the vector equation need be considered. Furthermore, in those accelerators which are characteristically electromagnetic (as opposed to electrothermal), the Lorentz force term $\mathbf{j} \times \mathbf{B}$ is usually dominant on the right side of the equation; it is this category which will, in fact, concern us in this chapter. Finally, it is frequently possible to classify a system as being in steady state, or else impulsively driven, in which case one or the other of the parenthesized terms in the convective derivative will predominate.

Thus, the equation of motion for steady-state acceleration of a plasma flowing in a rectangular channel is

$$\rho v_x \left(\partial v_x / \partial x \right) = j_y B_z - \partial p / \partial x, \quad (16.2.2)$$

assuming that the gas is well enough ionized so that the viscous drag against background neutrals can be neglected. Even this simple a formulation, when combined with equally simple assumptions about the energy equation and equation of state does not usually lead to an analytic solution for the flow; however, certain very special though unrealistic further assumptions (isothermal or adiabatic flows, etc.) can yield solutions which have instructive value.[3]

A second extreme case with which we will be mostly concerned in this section is that of the "plasma gun." Such devices are usually impulsive in operation, so that the Lorentz force overwhelms that due to pressure gradients. In addition, the plasma tends to be rather well localized and strongly accelerated, making the velocity time derivative the most important left-hand term. We are then left with

$$\rho \left(\partial v_x / \partial t \right) = j_y B_z + f_{vx}. \quad (16.2.3)$$

In plasma guns, one cannot generally neglect the viscous force term f_{vx}. This is because the usual mode of operation is one in which the accelerating body of plasma moves into a background of neutral propellant (usually inserted transiently or "puffed" into the channel prior to the current impulse), and entrains it into the moving system by ionization. This is the "snowplow" model. The magnitude of this drag is just the local volume

---

[3] G. W. Sutton and A. Sherman, "Engineering Magnetohydrodynamics." McGraw-Hill, New York, 1968.

16.2. ELECTROMAGNETIC PROPULSION

rate of mass entrainment multiplied by the velocity change of each new mass increment:

$$f_{vx} = -v_x (\partial \rho / \partial t). \qquad (16.2.4)$$

16.2.2. Impulsive Acceleration in a Linear Channel

16.2.2.1. Geometry. To specify the system under theoretical and experimental study here, we consider a simple linear accelerator consisting of two parallel plates (Fig. 1) connected on the left to a capacitive energy storage

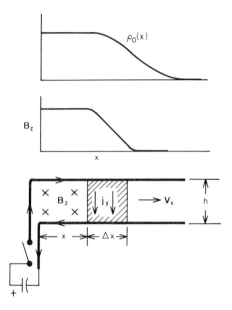

FIG. 1. Schematic diagram of linear impulsive accelerator.

and switch. The plates have a width w in the z direction (normal to the diagram) which is much larger than their spacing h. Further, we suppose that while the electrode plates truncate the system in the y direction, only spatial derivatives along z exist; thus, the problem is one-dimensional.

This "gun" is filled with some initial spatial distribution $\rho_0(x)$ of propellant gas, usually by a quick-opening value somewhere in the electrode region. The circuit switch is then closed, and a breakdown occurs in the propellant.

In general, the circuit inductance in these systems is small, and the current rise is very rapid, approaching 10^{12} amperes/sec in some instances. Conductivity of the plasma rises rapidly also, with the effect of forcing most of the current to flow in a thin layer at that boundary of the conducting gas nearest the energy supply. This current sheet is then forced, through the $j_y B_z$ interaction with its self-field, to move toward the muzzle of the gun along positive x. It is commonly observed that this magnetic piston entrains the $\rho_0(x)$ distribution of gas as it moves, and forces the mass out along x at high speed.

16.2.2.2. Diagnosis by Magnetic Probes. We now inquire as to the amount of information concerning the behavior of a plasma gun that can be obtained by mapping of the space and time distribution of the magnetic field. Magnetic probes are simple to employ, and when used in plasmas where perturbation effects are likely to be small, can give very precise results. The combination of density and temperature present in most plasma guns ($10^{16}/cm^3$ and 1–3 eV) is particularly favorable in that the heating rate of the probe surface is far too slow to result in significant evaporation of foreign material into the plasma in the usual impulse time; furthermore, the perturbation of plasma temperature outside the probe space-charge sheath is small.[4]

The usual probe is a small multiturn coil encased within an insulating jacket, typically 1–3 mm in diameter. The probe shaft is inserted into the accelerator vacuum chamber through appropriate seals, and is usually arranged to be movable in one or more coordinate directions. Output voltage from the loop, proportional to the time derivative of the component of **B** along the coil axis, is integrated and displayed on an oscilloscope whose sweep is triggered at the onset of current flow in the accelerator. Assuming reproducible behavior of the device from one shot to another, a set of photographic records of **B**(t) taken at a sequence of spatial positions can be cross-plotted as a family of **B**(r) distributions for different times. Through Maxwell's equations, one can infer the distribution of the current density as well as integral properties of the electric field.

In the particularly simple situation of a current sheet whose thickness Δx is much less than the accelerating channel length, we may use electrical circuit concepts to describe the behavior of the system. The two electrode plates, together with the plasma bridge between them become an inductor where

$$L = L'x, \qquad (16.2.5)$$

and

$$L' = \mu_0(h/w). \qquad (16.2.6)$$

---

[4] R. Lovberg, *in* "Plasma Diagnostic Techniques" (R. H. Huddlestone and S. L. Leonard, eds.), Chapter 3. Academic Press, New York, 1965.

We also suppose that the sheet has some net electrical resistance R.*
Now, the terminal voltage as seen by the energy supply is

$$V = IR + \partial \phi/\partial t \qquad (16.2.7a)$$
$$= IR + \partial(LI)/\partial t, \qquad (16.2.7b)$$

where we assume that inductance can be defined as

$$L = \phi/I. \qquad (16.2.8)$$

Then, the input power becomes

$$P = VI = I^2 R + IL\,(\partial I/\partial t) + I^2\,(\partial L/\partial t)$$
$$= I^2 R + \partial(\tfrac{1}{2}LI^2)/\partial t + \tfrac{1}{2}I^2\,(\partial L/\partial t). \qquad (16.2.9)$$

The three right-hand terms in Eq. (16.2.9) are respectively, resistance loss (to heating of the plasma), rate of magnetic field energy increase, and work done against the moving current sheet. When this equation is integrated over time, one obtains an expression for efficiency in a particular sense. The integral of the input power is clearly the total pulse energy; the integral of the second right-hand term is evidently zero and the remainder is

$$W_0 = \int_0^\infty I^2(R + \tfrac{1}{2}\,dL/dt)\,dt. \qquad (16.2.10)$$

Since the object of MHD propulsion is the direct transfer of momentum from electromagnetic field to plasma without heating, we see that an electrical efficiency can be defined by

$$\varepsilon_e \equiv \int_0^\infty (\tfrac{1}{2}I^2\,dL/dt)\,dt/W_0, \qquad (16.2.11)$$

where any resistive power, even though partially recoverable in a nozzle expansion, is considered here to be a loss.

Experimental determination of ε_e in an accelerator working in the current sheet limit requires only a measurement of the position x of the sheet as a function of time. Magnetic probes are ideal in this application, since a simple determination of the time and amplitude of the B_z jump

---

* While the use of cgs units has been adopted as standard for most work in plasma physics and also for this series, the very strong coupling between plasma physics and electrical engineering which characterizes the propulsion application, makes it most convenient (indeed, nearly necessary) to employ the mks system of electrical units here. The reader will find that converting units at the "interface" between the physics and engineering areas of the problem is, in the end, less clumsy than attempting to adhere to a consistent system of units throughout. One hopes that this dichotomy will be resolved in the future, but as of this writing, it has not.

which occurs when the current passes each probe location provides both a value of current I ($I = w \, \Delta B/\mu_0$), and of the sheet trajectory. Were the magnitude of ΔB not explicitly measured, I could be determined by a conventional measurement at the input terminals. We assume also, that W_0 is known.

The magnetic probe, provided that it can be traversed along y and z as well as along x, also enables the experimenter to verify his assumptions concerning the geometrical regularity of the plasma configuration. The simple procedure outlined above is only valid if the current is actually a thin sheet in the yz plane, occupying the full width w of the electrodes. Modest departures from this ideal will not invalidate the experiment, and it is found in practice that the planar ideal can be realized well enough for the kind of efficiency estimate we have described.

It is possible, of course, to evaluate the time history of the circuit inductance even if the discharge configuration is more complicated than a simple sheet; a three-dimensional mapping of the three field components is in principle enough to specify the inductance completely. However, a retreat to multidimensional mapping is always cumbersome, and at worst is impossible, this in the event that the configuration does not accurately reproduce itself from shot to shot of the discharge. It is unhappily true that most of the departures from simple current sheet structure in plasma guns are random in character, thus precluding the mapping procedure altogether. A stable operating mode is thus an absolute prerequisite for the useful employment of magnetic probes.

16.2.3. Experiment: Energy Inventory in a Coaxial Gun

An experiment conducted by Larson et al.[5] illustrates the usefulness of magnetic probing in efficiency measurements. Their accelerator, shown schematically in Fig. 2, employs a coaxial electrode pair having a 2:1 diameter ratio, with the propellant, nitrogen in this instance, injected into the interelectrode region through a system of ports in the inner electrode. Energy is stored in a specially fabricated capacitor which surrounds the system coaxially and which is connected to the electrode structure by a close-spaced pair of circular disks that comprise, as one wishes to look at it, a parallel plate transmission line, or a large diameter but exceedingly short coaxial line. No separate switch is employed in the circuit; rather, the injected propellant is its own switch, breaking down and conducting the discharge when its pressure becomes sufficiently high. This device is somewhat unique in its genre in that the circuit inductance

[5] A. V. Larson, T. J. Gooding, B. R. Hayworth, and D. E. T. F. Ashby, *AIAA J.* **3**, 977 (1965).

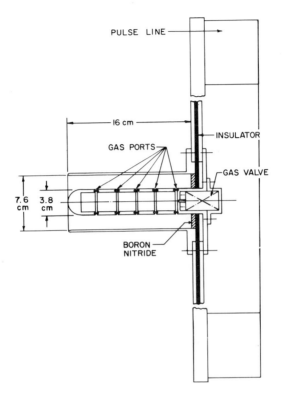

FIG. 2. Coaxial plasma gun of Larson et al. [A. V. Larson, T. J. Gooding, B. R. Hayworth, and D. E. T. F. Ashby, *AIAA J.* **3**, 977 (1965)]. (Courtesy of Convair Division of General Dynamics Corp.)

is so low that the storage capacitor no longer behaves like a simple capacitance, but rather assumes the character of a pulse-forming network. (Any capacitor will behave in this fashion if the oscillation period of the circuit into which it is connected is less than the propagation time of a pulse through the capacitor itself. Such an inequality is hardly ever achieved in energy storage systems, however.) As a result, the current quickly assumes a nearly constant value lasting for two pulse transit times of the capacitor, after which it decays fairly rapidly.

The storage capacitance of 22 μF is charged initially to 6.3 kV, with a maximum current of about 2×10^5 amperes flowing during the acceleration pulse.

Initially, in the sequence of probe measurements, B_θ (the only field component for an azimuthally uniform system) was sampled at different azimuths in order to ascertain whether some nonuniformity might exist; however, no azimuthal asymmetry could be detected.

Data were then accumulated over a single rz plane. Figure 3 is a set of oscilloscope trace photographs showing $B_\theta(t)$ at five different axial positions, all at a radius halfway between the electrodes. It is evident that the main radial discharge current flows in a thin layer which propagates at a fairly uniform velocity. It is remarkable that each of these photographs is an overlay of four traces from successive shots of the gun;

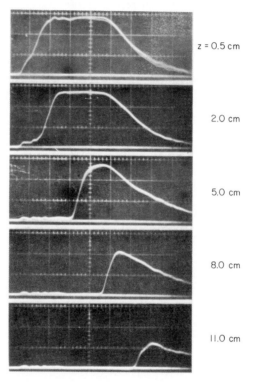

Fig. 3. Oscillograms of $B_\theta(t)$ at five axial positions in the coaxial gun. (Courtesy of Convair Division of General Dynamics Corp.)

in only the top set is there sufficient irreproducibility to separate the traces visibly.

Figure 4 displays the spatial distribution of B_θ at five different times. Here, even more clearly, one sees the propagation of a thin current sheet taking place. One concludes from these data that

1. The initial strong flow of current takes place near the insulating surface at the rear of the barrel ($z = 0$), and in a zone 2–3 cm thick.
2. After a few tenths of a microsecond, a well defined **current sheet** moves down the barrel at a velocity of 10^7 cm/sec.

16.2. ELECTROMAGNETIC PROPULSION

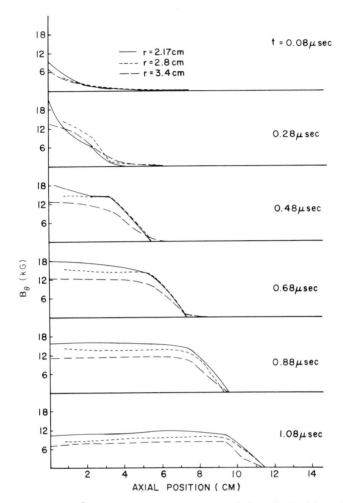

FIG. 4. Plots of $B^\theta(z)$ at five times. (Courtesy of Convair Division of General Dynamics Corp.)

3. The current flows almost entirely along r, with very little "tipping" of the sheet along z; this is evident because the three $B_\theta(z)$ plots for different radii all rise from zero to their maximum values in the same interval of z.

4. No significant current flow occurs away from the sheet. $\partial B_\theta/\partial z$ ($= \mu_0 j_r$) is nearly zero everywhere else, and B_θ varies closely enough as $1/r$ to imply negligible j_z.

It will be recalled from Eq. (16.2.9) that the input power VI at the accelerator terminals is partitioned between three terms: I^2R, the resistive heat loss; $d(\tfrac{1}{2}LI^2)/dt$, the rate of field energy increase, and $\tfrac{1}{2}I^2\,dL/dt$, the rate of work done on the moving piston. Larson et al. monitored V and I at the input terminals, and obtained L from Eq. (16.2.5), where here,

$$L' = (\mu_0/2\pi)\ln(b/a)$$
$$= 0.14 \quad \mu\text{H/meters}.$$

The partitioning of the integrated input power between field energy and work terms is shown in Fig. 5. The difference between the sum of

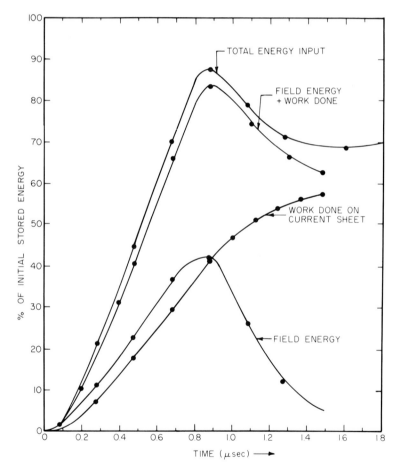

Fig. 5. Partitioning of energy in the coaxial gun. (Courtesy of Convair Division of General Dynamics Corp.)

these two terms (shown as a separate curve) and the total input energy is attributable to resistance.

Note that after about 0.8 μsec, the integrated input energy drops sharply, as does the field energy within the device. This occurs because a sudden voltage reversal (associated with the return of the reflected wave in the capacitor) has occurred at the terminals, and consequently, VI is negative and energy is being withdrawn from the accelerator. Nevertheless, I is still nonzero and so work is still being done on the current sheet.

When the acceleration cycle is complete, it is found that 69% of the initial stored energy (W_0) has remained beyond the input terminals, with 58% of W_0 identifiable as useful work. The resistive losses in this case are relatively small. Presumably, the energy withdrawn back into the capacitor after 0.8 μsec is still available for acceleration of another puff of gas.

The electrical efficiency ε, as defined in Eq. (16.2.11), indicates the fraction of input energy which has gone into the work term of Eq. (16.2.9). ε_e is not, however, descriptive of the overall efficiency of the accelerator in converting input power to directed kinetic energy of the exhaust, because a portion of the work term itself may go into heating of the propellant.

Consider the "snowplowing" of a neutral gas fill into the advancing current sheet. Irrespective of the detailed physical process of entrainment the very fact of entrainment means that a totally inelastic collision has occurred between each neutral atom and the sheet; the impact energy must be left internally within the plasma layer, probably as heat. In the special case of a current sheet moving at uniform speed into a uniform background density, exactly one-half of the work done by the magnetic field upon the sheet goes into internal energy and one-half into flow energy. Black and Jahn[6] have analyzed this situation for general $I(t)$ and $\rho_0(z)$. They find that the "dynamical efficiency," i.e., the ratio of final directed energy to total work is enhanced by (1) strong acceleration in the channel and (2) an initial filling density distribution which is peaked at the rear of the channel. One simple extreme example, the "slug model," in which all of the propellant initially resides in the sheet, can be seen to have unit dynamical efficiency, since there are no inelastic collisions at all to dissipate the available energy.

Recognizing the limit on dynamical efficiency imposed by their distributed mass injection and nearly constant sheet velocity, Larson *et al.* made measurements of the energy content of the exhaust by employing a calorimetoric plasma collector. This is simply a deep copper "bucket"

---

[6] N. A. Black and R. G. Jahn, *AIAA J.* **3**, 1209 (1965).

facing the gun muzzle and subtending a large enough angle to collect essentially all of the emerging plasma. Before it can escape from this collector, the plasma will have lost nearly all of its energy in inelastic collisions with the walls; this will include ionization energy, since wall recombination is highly probable. The calorimeter is suspended in the vacuum system on supports of very low thermal conductance; hence, each discharge of the accelerator provides a step increase in the collector temperature, which does not decay significantly between shots. An exhaust energy average, as well as a relatively large and accurately measurable temperature change can then be obtained if many consecutive discharges are made into the collector.

It may be argued that the calorimeter measures the internal as well as translational energy of the beam, and is consequently of doubtful value in estimating overall efficiency. While this is true in principle, it turns out that in heavy-ion plasmas having directed kinetic energies in the range of hundreds of electron volts, the energy remaining in the stream after it has left the accelerator and entered the calorimeter is almost entirely translational. The reason is that while a freshly entrained ion may have a large thermal kinetic energy within the sheet, it rapidly loses it to the background electrons, which in turn lose it rapidly through radiative processes. (In the coaxial gun experiment of Larson, the ion-electron equipartition time was estimated to be about 0.25 μsec for 5 eV electrons.) While it is true that the calorimeter will itself intercept a fraction of the radiative loss, this fraction will usually be small, since the radiation is isotropic.

In the experiment described here, the actual recovery of energy achieved in the calorimeter was about 45% of W_0, this after optimization of the accelerator electrode dimensions. The authors observe that 50% is the maximum which should theoretically have been possible, given their sheet velocity and gas fill conditions.

16.2.4. The Problem of Distributed Currents

We have seen that a pulsed accelerator in which the plasma current flows in a well-localized sheet enables the experimenter to employ circuit element approximations, and thus to estimate, from magnetic field data alone, the partitioning of energy between resistive loss and $\mathbf{F} \cdot \mathbf{v}$ work. The current sheet idealization, while useful when conditions permit its use, fails when current in the plasma is distributed over a region comparable in extent to the accelerator dimensions. Specifically, we find that for broadly distributed currents, a simple mapping of the space and time distribution of the magnetic field no longer suffices to distinguish between resistive heat loss and $\mathbf{F} \cdot \mathbf{v}$ work. We may see the reason for this in either of two ways.

First, we may consider the simple Ohm's law for a moving medium,

$$\mathbf{j} = \sigma(\mathbf{E} + \mathbf{v} \times \mathbf{B}), \tag{16.2.12}$$

where we suppose for now that the conductivity σ is scalar. Solving for $\mathbf{E}$, and multiplying by $\mathbf{j}$ to get the local power input to the medium from the field, we obtain

$$\begin{aligned}\mathbf{j} \cdot \mathbf{E} &= j^2/\sigma - \mathbf{j} \cdot (\mathbf{v} \times \mathbf{B}) \\ &= j^2/\sigma + \mathbf{v} \cdot (\mathbf{j} \times \mathbf{B}),\end{aligned} \tag{16.2.13}$$

where again, we have separate right-hand terms which are resistive heating and $\mathbf{F} \cdot \mathbf{v}$, respectively. This time, however, the terms are point functions and so must be integrated over both space and time in order to yield an efficiency figure (ε_e). The integral of the left-hand side is still the total input energy at the terminals, and so is known, as before. However, before either of the right-hand integrals can be evaluated (one is sufficient to determine ε_e), we must know the spatial distribution of either the conductivity σ or the plasma velocity vector $\mathbf{v}$, and a magnetic probe cannot provide either one. The probe data will, in general, show the experimenter that the magnetic field system is advancing along the accelerating channel, but unless the distribution terminates abruptly at the front (as in a current sheet) it is impossible to tell whether the advance is due to diffusion through the plasma (resistive) or to motion of the conductor into which the field is frozen (work).

The very concept of inductance loses much of its precision when one is dealing with distributed currents. In deriving Eq. (16.2.9), we defined inductance as

$$L = \phi/I,$$

where ϕ is the total circuit flux, while at the same time we assumed the definition

$$L = 2W/I^2,$$

where W is the energy in the magnetic field. Suppose one were to divide the total flux in the rectangular system of Fig. 1,

$$\phi = h \int_0^\infty B_z(x)\, dx,$$

by the input current, and compare this inductance with that obtained by dividing twice the field energy

$$W = hW \int_0^\infty (B^2/2\mu_0)\, dx \qquad (w \gg h)$$

by the square of the current. It will be discovered that except for the special case of a perfectly thin current sheet bounding B_z in the yz plane, the two definitions are contradictory. The reader may easily generate simple trial cases to verify the contradition; for the simple instance of a uniform j_y between $x = 0$ and some $x = a$, it will be found that these two definitions of L differ by a factor of $\frac{3}{2}$. This point has not been totally ignored in basic texts on electromagnetic theory but is not usually made very clear. Grover[7] in his classic treatise on inductance calculations, points out the difficulty by showing that the notion of circuit flux is not well defined if currents flow in a conductor of greater than zero cross-section. Energy is still well defined, however, and so one should always regard the inductance as relating current and field energy. Grover uses the concept of "partial flux linkages" to resolve the dilemma; however, when such procedures become necessary, it is clear that the utility of circuit concepts is lost, and one can see the problem in a simpler and more fundamental way by returning to the field equations.

16.2.5. Additional Diagnostics

It can be seen from the foregoing that whether the accelerator under study is being analyzed in the current sheet limit, or as a distributed current system, the information obtainable from magnetic field mapping is ambiguous. A sensor which would measure local plasma velocity with good space and time resolution (in the scale of a pulsed plasma gun) would, of course, remove the ambiguity, and would, indeed, make other plasma diagnostic techniques unnecessary, since what the investigator ultimately wants is the complete velocity distribution of the exhaust. Unfortunately, no such sensor has been available, although certain closely related devices have been useful.* Several workers[8,9] have employed piezoelectric pressure transducers to measure the instantaneous momentum flux at points within the moving plasma. The extraction of velocity data from momentum obviously requires a knowledge of density, and density is a parameter amenable to measurement, although not usually with the spatial and temporal resolution of which the momentum and field probes are capable.

Particle analyzers such as biased Faraday cups or curved plate and magnetic spectrometers are unsuitable for measurements within pulsed

---

[7] F. W. Grover, "Inductance Calculations." Van Nostrand, Princeton, New Jersey, 1946.

[8] M. O. Stern and E. N. Dacus, *Rev. Sci. Instrum.* **32**, 140 (1961).

[9] I. R. Jones, Aerospace Corp. Rep. No. TDR-594(1208-01)TR-3. Aerospace Corp., Los Angeles, California, 1961.

---

* In Part 16.3, velocity is measured by the use of Langmuir probes. However, special plasma conditions, not achieved in coaxial guns, are necessary for this procedure to work.

accelerators, because they are always too bulky, and also become inoperative if the plasma incident upon them is too dense. A necessary operation upon any plasma whose ion velocity is to be measured by electromagnetic means is the separation of the ions from their companion electrons. However, this is only feasible if the plasma is sufficiently tenuous that the Debye length is greater than the scale size of the electrode system doing the stripping, and such a condition is hardly ever achieved except in the well-expanded exhaust at a distance from the accelerator muzzle. In this region, however, some excellent beam analysis has been done.[10]

16.3. Electric Propulsion as a Plasma Physics Problem: The Magnetoplasmadynamic Arc

16.3.1. The MPD Arc—General Properties

During 1964, in the course of a series of investigations of a steady-state electrothermal arc-jet device, it was discovered by Ducati[11] that by considerably lowering the rate of propellant inflow below what had been considered standard, a qualitatively new mode of arc behavior was produced in which electrode erosion rates became extremely low, and in which, even more importantly, the overall propulsive efficiency of the device became high, approaching 50%. Widespread activity was stimulated by this finding, with most workers verifying the initial results.

Figure 6 schematically depicts a typical magnetoplasmadynamic arc or MPD arc as the device has become known. A nozzle-shaped outer

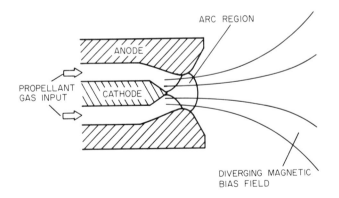

Fig. 6. Schematic diagram of MPD arc.

[10] L. F. C. Liebing, *AIAA J.* **6**, 1763 (1968).
[11] A. C. Ducati, G. M. Giannini, and E. Muehlberger, *AIAA J.* **2**, 1452 (1964).

anode, usually water-cooled, contains within it along its axis a tungsten cathode whose conical tip is usually placed slightly behind the nozzle throat. Propellant gas flows through the annular gap between electrodes, and out into the exterior vacuum. An arc discharge is struck between the electrodes, tending to localize in the throat region. It ionizes, heats, and expels the propellant as a plasma.

It was found, immediately subsequent to the first experiments that an axial magnetic bias field set up by a coil just over or slightly behind the discharge tended to stabilize the arc performance, and so this additional feature is by general agreement incorporated into the definition of an MPD arc.

Parameters of a typical arc are[12]: throat diameter, 1 cm; arc voltage, 35 volts; arc current, 2500 amperes; bias field, 2000 gauss; efficiency, 35%; propellant, ammonia; propellant flow rate, 0.015 gm/sec; exhaust velocity,* 6.4×10^4 meters/sec.

For a long period, most of the research on these arcs was done by the use of external diagnostics such as measurements of thrust, arc voltage and current, mass flow rate, and external exhaust measurements. The interior of a typical MPD arc is as inconvenient and hostile an environment for detailed plasma measurements as can be imagined. A steady-state power density of about 10^5 watts/cm^3, together with a very confined and nearly inaccessible arc chamber, make conventional probing essentially impossible; in such an environment, any probe would be destroyed in seconds. Nevertheless, the quite remarkable character of this class of arc has made it a most interesting object for direct internal examination.

16.3.2. The Pulsed MPD Arc

One evident possibility for easing the difficulty of probing would be to pulse the arc, i.e., to turn it on only long enough to allow the attainment of steady state. If this period is sufficiently short, one would be able to obtain useful data from probe-type sensors before they are heated to destruction. Unfortunately, this is not practical in the standard MPD arc, because a period of several seconds is usually required between turn on and the achievement of steady operation. Apparently, the delay time is

---

[12] S. Bennett, Avco Space Systems Div. Rep. No. RAD–TR–65–37, NASA CR54867m. Avco Space Systems, Wilmington, Massachusetts, December 1965.

---

* Propellant velocity here is an equivalent velocity, the ratio of thrust to mass flow rate. The engineering literature employs the notion of "specific impulse," I_{sp}, where

$$I_{sp} \equiv \mathbf{u}_e/g = T/\dot{m} \text{ (sec)},$$

with T in pounds and $\dot{m}$ in pounds per second.

16.3. ELECTRIC PROPULSION AS A PLASMA PHYSICS PROBLEM

that required by the cathode to heat, under ion bombardment, to thermionic emission temperature. The successful operation of a pulsed arc would then seem to require either (1) the use of a cathode of such low heat capacity (probably a thin shell) that its self-heating time would be very short or (2) the use of some preheating technique for the cathode, which would allow steady state almost instantly after arc initiation.

The device described here was built according to the second prescription, i.e., it employs a preheated cathode. It is also larger than conventional arcs, and simpler in configuration, in order that probing of the active arc region might be done easily and with spatial resolution significantly finer than the interelectrode distance.[13]

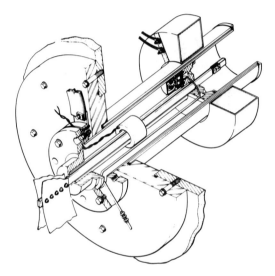

FIG. 7. The pulsed MPD arc of Kribel *et al.* [R. Kribel, C. Eckdahl, and R. Lovberg, *AIAA J.* to be published].

Figure 7 is a cutaway drawing of the apparatus. The cathode is a single "hairpin" turn of tungsten ribbon, preheated in 0.3 sec to about 3300° K by a pulse of 60 Hz alternating current. The anode is a cylindrical shell of stainless steel 9 cm in diameter. Propellant, argon in this case, is pulsed into the accelerator by a fast-opening solenoid valve and nozzle system located in the ceramic insulator at the rear of the arc chamber. The insulator also serves as a vacuum wall. An external coil, energized by a pulse of current from a bank of capacitors, supplies the magnetic bias field.

[13] R. Kribel, C. Eckdahl, and R. Lovberg, *AIAA J.* to be published.

Operation of the system involves the following sequence of events:
(1) The cathode heating pulse is initiated.
(2) At 0.3 sec, the heating pulse is turned off, gas injected, and the bias field pulsed on.
(3) When the bias field reaches maximum strength, and the injected propellant is flowing steadily past the cathode (both take about 2×10^{-3} sec), the main arc pulse is turned on; it is approximately a constant current step lasting 500 μsec, obtained from a pulse-forming network.

Conditions adopted as standard for these experiments are: arc voltage, 75 volts; arc current, 550 amperes; bias field (at arc), 500 gauss; argon flow rate, 0.02 gm/sec. These conditions are evidently different from those quoted earlier for a steady-state arc, and the system geometry is a drastic variation from "normal." It is reasonable to question whether one is actually approximating a typical MPD system, or whether the device is a distinctly new species. It has turned out, however, that nearly all of the qualitative operating characteristics of ordinary MPD arcs have been duplicated by the pulsed unit.

It can be seen, by multiplying voltage, current, and pulse length, that the total energy consumed in each pulse is only about 20 joules; of this, only a very small fraction, probably less than 10^{-2}, is intercepted by a probe, and so, probe erosion is entirely negligible. Yet it is apparent from all of the diagnostic data that a steady state is achieved in a time very small compared to 500 μsec.

A further distinct advantage of pulsed operation is that whereas in a dc arc, the normal propellant flow rate is so large as to prevent any ordinary vacuum pump from achieving better than several torr background pressure, the propellant pulse is so short in the present device that no significant flow from the far end of the vacuum tank occurs before the arc pulse is terminated. Thus, a good approximation to the hard vacuum of a space environment is presented to the accelerator.

16.3.3. Experiments with the Pulsed MPD Arc

16.3.3.1. Magnetic Measurements.

16.3.3.1.1. THE USE OF ROGOWSKI COILS. The problem of determining plasma current distributions in the MPD arc contrasts interestingly with the same measurement as done in a fast plasma gun system (Section 16.2.2.2). In the previous example, the entire magnetic field was set up by plasma currents, and so, spatial variations in the quite strong field structure were of the same order as the field itself. In the MPD arc, on the other

16.3. ELECTRIC PROPULSION AS A PLASMA PHYSICS PROBLEM

hand, self-fields from plasma currents are much smaller, usually by two orders of magnitude, than the bias field, and also smaller by about the same factor than those in plasma guns.

It is very convenient in this situation to make current density measurements with a current-transformer probe known as a Rogowski coil, rather than employing conventional loop probes.

The Rogowski coil is a small, multiturn solenoid wrapped on a toroidal form, as shown schematically in Fig. 8. It has the very useful property that its output voltage is proportional to the time derivative of the total current passing through its open aperture, irrespective of the spatial

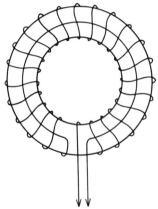

FIG. 8. The Rogowski coil configuration. Ideally, output voltage results only from a time-varying current through its aperture.

distribution of that current. This property is made plausible by the following argument: The total voltage being the sum of voltages produced by the separate turns, we may integrate along the toroidal (minor) axis to obtain

$$V = \oint \dot{B} A \, dn$$
$$= (NA/l) \oint \dot{B} \, dl$$
$$= n A \mu_0 \dot{I}/l, \qquad (16.3.1)$$

where A is the cross-sectional area of the torus, l is its length, and N the total number of turns, assumed uniformly distributed. The last equality follows from Ampére's law. With integration by a passive RC network having $\tau = RC$ much larger than the observation interval, we finally have

$$V = \mu_0 N A I / R C l. \qquad (16.3.2)$$

While it has been assumed that the coil has a uniform cross section and turn density, it has not been assumed to be a circle. Indeed it may have any shape of open aperture, provided only that it actually closes upon itself.

One may note in Fig. 8 that the wire connecting the last turn to the external circuit is brought back along the full length of the coil to join the second lead. This prevents the coil as a whole from acting as a single-turn loop, sensitive to magnetic flux, rather than current, which might also pass through its aperture. Equation (16.3.2), while showing that the coil output is independent of the exact distribution of current through it, also implies an even more important property: the coil output will be zero if no current links it, and this will be true even when it is immersed in strong, fluctuating fields. Herein lies its very great utility for measurements in systems such as the MPD arc: weak currents, whose self-fields are much smaller than externally imposed fields, may be quite easily detected, where their measurement with conventional probes might be nearly impossible.

The coordinate system employed for all measurements in this system is shown in Fig. 9. Included also in this figure are arrows representing the strength of the z component of bias field for "standard" arc conditions.

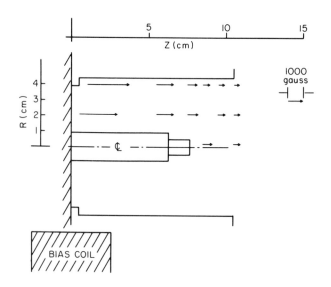

Fig. 9. The coordinate system used for the pulsed MPD arc experiments is shown by scales adjacent to the accelerator outline. Arrows indicate strength of the axial component of bias field within the system.

16.3. ELECTRIC PROPULSION AS A PLASMA PHYSICS PROBLEM

16.3.3.1.2. Distribution of Current. First measurements of the current flow pattern in the MPD arc were attempted with magnetic loop probes oriented to couple B_θ, the azimuthal field component associated with radial current flow between the coaxial electrodes. Since the externally imposed field possessed only r and z components, one could avoid coupling it into the B_θ sensor by very careful alignment.

Early data from the probe are shown in Fig. 10. The radial position common to both was $r = 2$ cm (anode radius, 5 cm) and the two axial positions were both behind the cathode ($z = 7$ cm). Both traces have the shape of the main current pulse, about 500 μsec long; superimposed upon

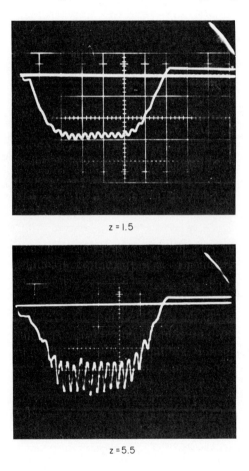

Fig. 10. Oscillograms of $B_\theta(t)$ at $z = 1.5$ cm and $z = 5.5$ cm. Sweep rate is 100 μsec/cm. The oscillating component of the current pulse becomes markedly stronger near the cathode suggesting an approach toward a rotating spoke.

both, however, is a high-frequency oscillation which is distinctly stronger for the position near the cathode.

It was inferred from these data that the main radial current flow, which was expected to occur near the cathode, and which was also expected to have good azimuthal symmetry (on the basis of previous published observations) is not symmetrical, but rather, possesses strong azimuthal irregularities which rotate about the axis at a frequency of about 30 kHz.

At this point, a Rogowski coil having an aperture diameter of approximately 2 cm was inserted into the system, and a traverse made along z at a radius of 3.5 cm. The integrated coil out put shown in Fig. 11. Here,

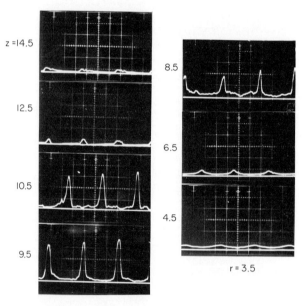

FIG. 11. Integrated signals from a Rogowski coil traversing the active arc region; sweep, 10 μsec/cm. The current plainly flows in a channel which periodically threads the coil, and which is confined near $z = 10$ cm.

the character of the current channel becomes very clear. It has the form, at this radius, of a highly localized spoke centered in its axial distribution near the anode lip at $z = 10$ cm, and occupying an azimuthal interval of about 40°. The direction of rotation is along $\mathbf{j} \times \mathbf{B}$, the direction of "swirl," in MPD arc parlance.

More complete Rogowski coil surveys, employing three-dimensional mappings with coils oriented along each of the principal cylindrical coordinate axes, have yielded the three-dimensional current channel distribution shown in Fig. 12.

16.3. ELECTRIC PROPULSION AS A PLASMA PHYSICS PROBLEM

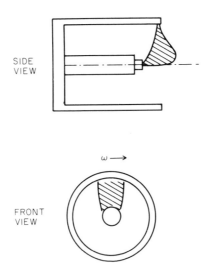

FIG. 12. The approximate current channel shape inferred from Rogowski coil data.

16.3.3.1.3. SPATIAL MAPPING ALONG θ. Since the current channel rotates about the system axis, the problem of determining spoke structure in the θ direction becomes complicated. However, it was found that with the rotation rate of the arc quite uniform, at least during the steady portion of the current pulse, it was possible simply to convert the time variation of sensor output into azimuth.

A trigger signal generator, in this case an optical detector focused upon a selected point on the anode lip, provided a sweep trigger to the display oscilloscope each time the spoke crossed that particular azimuth; the sweep duration was slightly in excess of one rotation period. The scope beam itself was turned on only during 300 μsec of steady arc current. Thus, an overlay of up to ten traces was obtained on each photograph for a single arc pulse, the time axis corresponding directly to azimuth, and the scatter between the several traces in the overlay giving a display of the error to be attributed to the particular measurement due to irreproducibility.

With the θ traverse provided automatically, it was only necessary in this experiment to traverse the probes along r and z.

16.3.3.2. Electric Field Measurement, and Inferences from Ohm's Law.

16.3.3.2.1. THE GENERALIZED OHM'S LAW. If one writes separate equations of motion for the electron and ion components of a plasma and combines them, assuming quasi–neutrality and scalar pressure, as well as neglecting certain terms which depend upon the electron–ion mass ratio, one obtains, upon adding the two equations, the MHD equation of motion for the plasma as a whole, and upon subtracting, a relation known as the generalized Ohm's law. This derivation is carried out in several standard works,[14–16] and we will not reproduce it here.

A form of Ohm's law suitable for the problems we will encounter in this work is

$$\mathbf{E} + \mathbf{v} \times \mathbf{B} - (1/n_e e)[\mathbf{j} \times \mathbf{B} - \nabla p_e] - \eta \mathbf{j} = 0, \quad (16.3.3)$$

where

$$\mathbf{v} = (m_e \mathbf{v}_e + m_i \mathbf{v}_i)/(m_e + m_i) \quad (\simeq \mathbf{v}_i),$$

$$p_e = nkT_e,$$

and

$$\eta \equiv m_e/n_e e^2 \tau_e,$$

where τ_e is the mean time between collisions of electrons with either ions or neutrals.

The importance of the generalized Ohm's law for our present purposes in this chapter is twofold. First, we will use it to enter into the realm of plasma physics, as distinct from magnetohydrodynamics, since (16.3.3) rests on the distinctiveness of ions and electrons. Second, it affords a means of inferring the distributions of some quantities of interest (e.g., electron density or plasma velocity) from other directly measured parameters such as $\mathbf{E}, \mathbf{B}$, etc.*

In the present experiment Ohm's law has been employed in dealing with the following problems:

(1) Since the current is observed to be confined to a constricted, rotating channel, the question arises as to whether the plasma itself

---

[14] L. Spitzer, Jr., "Physics of Fully Ionized Gases," 2nd ed., p. 23. Wiley (Interscience), New York, 1962.

[15] W. B. Thompson, "An Introduction to Plasma Physics," p. 223. Addison-Wesley, Reading, Massachusetts, 1962.

[16] B. S. Tanenbaum, "Plasma Physics," p. 140. McGraw-Hill, New York, 1967.

---

* The distinction between MHD and plasma physics may also be made by slightly rewriting Eq. (16.3.3) as

$$\mathbf{j} = \sigma_0(\mathbf{E} + \mathbf{v} \times \mathbf{B}) - (\Omega/B)(\mathbf{j} \times \mathbf{B} - \nabla p_e), \quad (16.3.4)$$

where $\Omega \equiv \omega_e \tau_e$ and where ω_e is the electron gyrofrequency. Magnetohydrodynamics assumes $\Omega = 0$, i.e., that the conductivity is scalar. However, if $\Omega > 0$, the conductivity becomes a tensor quantity, which means that in a frame of reference co-moving with the fluid, $\mathbf{E}$ and $\mathbf{j}$ are, in general, not parallel. Such behavior is characteristically a plasma property, and can only be understood by considering the detailed motion of charge carriers.

behaves in this fashion, i.e., whether the plasma actually possesses an azimuthal velocity component

$$v_\theta = r\omega,$$

where ω is the current spoke angular frequency, or whether conversely, the current column migrates through a relatively slower spinning plasma.

(2) The plasma is caused to swirl in the accelerator nozzle by the Lorentz force

$$F_\theta = j_r B_z - j_z B_r;$$

accordingly, a radial acceleration is applied to the plasma, and if some countervailing confinement force is not applied, a substantial loss of plasma and nonpropulsive energy will occur through radial outflow. It is thus important to ascertain the radial and axial components of velocity as well as the azimuthal component.

The procedure followed in this experiment was to measure directly the quantities

$$\mathbf{B}(\mathbf{r}, t), \quad \mathbf{J}(\mathbf{r}, t), \quad \mathbf{E}(\mathbf{r}, t), \quad n_e \mathbf{v}(\mathbf{r}, t).$$

Advantage was taken of the fact that along certain coordinate directions and in certain spatial locations, one or more terms in the full Ohm's law vanish, allowing solution for parameters which could not be uniquely specified if all terms were retained.

In Sections 16.3.3.2.2 and 16.3.3.2.3 we discuss inferences which can be made from electric field measurements (in addition to $\mathbf{B}$ and $\mathbf{j}$) and in Section 16.3.3.3, the incorporation of ion flux measurements into the program is shown to result in a complete specification of $\mathbf{v}$ as well as of plasma resistivity (3.3.4).

16.3.3.2.2. MEASUREMENT OF E. The electric field in plasma of the kind produced by accelerators such as the MPD arc (or higher power pulsed devices) is usually inferred from measurements of electric probe floating potentials. The floating potential V_f is that value attained by an electrode, immersed in the plasma, into which no net current flows. It is negative with respect to plasma space potential V by a small multiple of kT_e/e, i.e.,

$$V = V - \alpha k T_e/e, \qquad (16.3.5)$$

where α is typically between 3 and 4. Thus,

$$\mathbf{E} = -\nabla V = -(\nabla V_f - \alpha k \, \nabla T_e/e), \qquad (16.3.6)$$

and we see that $\mathbf{E}$ may be derived from floating potential measurements provided (1) that $T_e(\mathbf{r})$ is known, or (2) that it can safely be assumed that $\alpha k \, \nabla T_e/e$ is negligible compared to $\mathbf{E}$. The current drawn through the probe by the potential measuring instrument must also be negligible compared to the available random electron current in order that the assumption of floating potential may be valid.

In the present experiment, V_f was first mapped with a single floating probe. Subsequently, a closely spaced probe pair was used to infer directly the average component of **E** along the direction of separation.

Figure 13 is a floating potential contour map, with the distributions at two opposite azimuths shown in this section. The spoke is at the azimuth

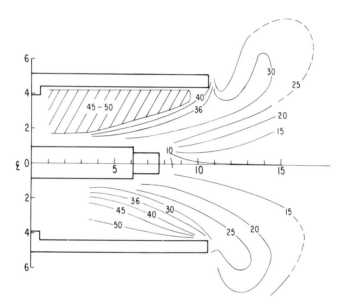

Fig. 13. Contours of equal floating potential. The current spoke lies in the upper half plane. Cathode ref.: PL, 1200 V; BF, 900 V; $V_A - V_k \approx 70$ V.

of the upper half-plane. The equipotential lines are very closely parallel to the magnetic bias field set up by a coil centered at about $z = 0$, just outside the anode cylinder. The total voltage between electrodes is 70 volts; it can be seen in the figure that no point in the field is at a floating potential higher than about 50 volts with respect to the cathode, which implies the existence of an anode sheath drop of over 20 volts. (The electron temperature in this plasma was found spectroscopically to be about 1.1 eV, so the probe sheath correction to the indicated potential was not large here.)

16.3.3.2.3. INFERENCE OF v_θ. The possibility of extracting velocity information at this point rests upon the possibility of simplifying Ohm's law in applying it to certain positions in the plasma. In the experiments

16.3. ELECTRIC PROPULSION AS A PLASMA PHYSICS PROBLEM

described here, the region near $r = 3.0$ cm, $z = 10.0$ cm (Figs. 9 and 12) was selected as being representative of the spoke. It has the further significant merits in terms of simplified treatment:

(1) j_θ is entirely negligible.

(2) Spectroscopic measurements indicate that $\nabla p_e/n_e e$ is negligible, except for its azimuthal component, which itself is relatively small. (We will not discuss the optical measurements in detail here, but refer the reader to the paper by Kribel et al.[13])

(3) B_θ is negligible in comparison with B_r and B_z.

Thus, we may reduce Eq. (16.3.3) to three much simpler component equations:

$$E_r = \eta j_r - v_\theta B_z, \tag{16.3.7}$$

$$E_z = \eta j_z + v_\theta B_r, \tag{16.3.8}$$

$$E_\theta = v_r B_z - v_z B_r + (1/n_e e)[j_z B_r - j_r B_z - \nabla_\theta p_e]. \tag{16.3.9}$$

Data obtained from the Rogowski coil survey and from a differential

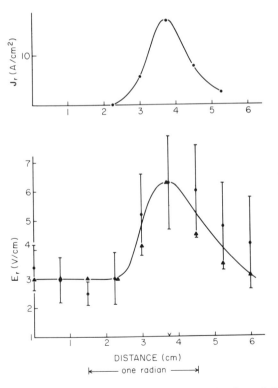

FIG. 14. Distributions of radial components of current density and electric field.

floating probe measurement of E_r may be compared in Fig. 14. (Maps of the bias field distribution were made prior to actual firing of the arc, and were assumed to apply to the system with the arc on.)

Here, the azimuthal distributions of both j_r and E_r are shown. The error bars on E_r arise from the cycle-to-cycle scatter as the spoke swept past the electric probe several times during a single shot of the arc. "Distance" here refers to circumferential displacement at $r = 3$ cm.

It is seen that E_r peaks at the center of the current spoke. However, it assumes a nearly steady nonzero value at azimuths where j_r has vanished. We may thus employ an even more compressed version of Eq. (16.3.7), which is now

$$E_r = v_\theta B_z, \qquad (16.3.10)$$

to obtain a value for the plasma swirl velocity in this region. This value agrees well with the observed spoke velocity. In Fig. 14, the triangular points in the circumference scale interval between 0 and 2 cm are values of E_r *defined* by the assumption that $E_r = \omega r B_z$. Equally good agreement between spoke velocity and E_z/B_r is obtained for the zero current region.

In Fig. 14 again, the triangular points in the spoke region of the E_r plot are a fit achieved by assuming a certain single value of η for the region. The relation of this resistivity to plasma temperature will be discussed later.

On the basis of this excellent agreement between inferred v_θ just outside the spoke azimuth and the observed spoke velocity, it is concluded that the plasma within the current channel is actually co-rotating with it.

16.3.3.3. Ion Flux Measurements.

16.3.3.3.1. LANGMUIR PROBE APPLICATION. It is possible, in a plasma of the kind produced by the MHD arc, to use Langmuir probes to obtain ion velocity information in a relatively simple way. This is made possible by the following fortunate combination of plasma parameters:

(1) The Bohm velocity

$$v_B = (kT_e/m_i)^{1/2}, \qquad (16.3.11)$$

which governs the ion flux to a negatively biased probe when $T_i < T_e$, is in this case very small ($\sim 2 \times 10^5$ cm/sec^1) compared with the mean ion speed; thus, the ion velocity itself determines ion current.[17]

(2) The ion-ion mean free path is large compared to probe dimensions. The probe does not, then, produce any significant aerodynamic perturbation upon the ion flow.

---

[17] D. Bohm, *in* "The Characteristics of Electrical Discharges in Magnetic Fields" (A. Guthrie and R. K. Wakerling, ed.), Chapter 3. McGraw-Hill, New York, 1949.

(3) The ion gyroradius is much larger than probe dimensions, which allows neglect of the effects of the magnetic field on ion collection.

(4) The Debye length is about 5×10^{-4} cm in this plasma making the sheath far smaller than probe dimensions, and rendering orbital effects negligible.

Under these conditions, we may employ the simplest Langmuir theory with confidence. If sufficient negative bias is applied to repel essentially all electrons, we can treat the probe as a simple geometrical interceptor of ions, i.e., we may suppose that the probe collects all the ions that would have passed its position had it not been in place.

Suppose, then, that under these conditions one positions a planar probe so that it faces the negative x direction. It then collects only ions with a component of velocity toward positive x. Its surface current density is

$$j_{i+} = Ze \int_0^\infty u_x f(u_x) \, du_x, \qquad (16.3.12)$$

where $f(u_x)$ is the ion velocity distribution in one dimension, and Z is the average ionization level.

This may be written as

$$j_{i+} = Ze \left[\int_{-\infty}^\infty u_x f(u_x) \, du_x - \int_{-\infty}^0 u_x f(u_x) \, du_x \right] = Zen_i u_x + j_{i-}, \qquad (16.3.13)$$

where j_{i-} is the current density collected by a probe electrode facing toward the positive x axis.

For a stationary plasma, $u_x \, (= v_x) = 0$ by definition, and so the ion currents j_{i+} and j_{i-} collected in the two opposite directions will be equal. Conversely, an observed inequality in the two current components implies the existence of a plasma drift velocity given by

$$v_x = (j_{i+} - j_{i-})/Zen_i. \qquad (16.3.14)$$

Thus, plasma motions may be inferred from the differential current between oppositely facing planar probes.

16.3.3.3.2. v_i AND T_e ESTIMATES. For the present experiment, the two probe electrodes were platinum tabs approximately 2-mm square mounted on opposite faces of a rectangular support shaft of about the same size. Platinum has several advantages for use in electrodes, the most important being that the secondary electron emission coefficient for incident argon atoms and ions is less than 5% for the energies encountered here; also, the effect of surface contamination appears negligible.

Each electrode was connected to a bias circuit as shown in Fig. 15. The large capacitor provides a low power supply impedance for surge currents.

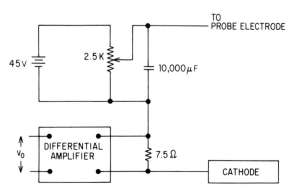

FIG. 15. Bias circuit for electric probe.

Initially, the probe was positioned at $z = 10$ cm, the position of the anode lip, with the electrodes facing in the $+\theta$ and $-\theta$ directions. Currents from both electrodes were recorded at each of several radii using the multi-overlay technique described in Section 16.3.3.1.3. The reason for beginning with azimuthal flux measurements is that an independent

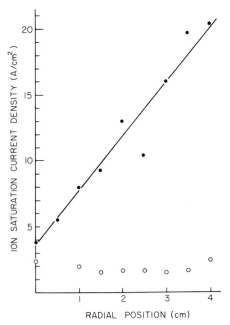

FIG. 16. Peak ion current versus radial position for two planar probes facing $+\theta$ and $-\theta$ respectively. (●) Front; (○) rear.

16.3. ELECTRIC PROPULSION AS A PLASMA PHYSICS PROBLEM

estimate of v_θ was available (Section 16.3.3.2.3), so that it was in this case possible to infer a value for the ion density itself.

Figure 16 is a plot of the peak ion current density for each electrode (which occurred as the spoke crossed the electrode on each revolution) as a function of probe radius. The current to the electrode facing the oncoming spoke is seen to be a nearly linear function of radius. This implies a nearly constant ion density over the radial interval, since for the rigid spoke $v_\theta = \omega R$. The density inferred from these data turns out to be

$$n_i = 1.0 \times 10^{14}/\text{cm}^3,$$

assuming negligible second ionization.

The current to the electrode facing the direction of the receding spoke is seen to behave quite differently. There is essentially no variation of ion current with radius. We may first venture the guess that the number of ions in the rotating spoke having a backward component of u_θ is negligible. This is equivalent to assuming that ω_r greatly exceeds the ion thermal speed; we will see later that this assumption is valid. Under such a condition, the ion flux into the rearward-looking electrode must be governed by the Bohm criterion, so that the saturation current is determined by electron temperature and ion density, neither of which, as will appear below, is a strong function of radius.

We may employ the above hypothesis to arrive at an estimate of electron temperature. It was shown before that the forward-looking electrode will receive a current density

$$j_{i+} = Zen_i u_\theta.$$

The Bohm current is, closely,

$$j_B = 0.5 Z n_i e (kT_e/m_i)^{1/2}. \tag{16.3.15}$$

Thus,

$$kT_e = m_i (2j_B v_\theta/j_{i+})^2. \tag{16.3.16}$$

At $r = 3$ cm, $v_\theta = 7.5 \times 10^5$ cm/sec^1. Also, $j_{i+}/j_B \simeq 10$, from Fig. 16. These numbers yield

$$kT_e = 1.0 \text{ eV}.$$

A spectroscopic determination of kT_e, made by intercomparing the intensity ratios of sixteen AII lines, gave

$$kT_e = 1.1 \pm 0.3 \text{ eV},$$

and also showed the electron temperature to be nearly uniform over the spoke. It is concluded, then, from these data that n_i does not vary appreciably with radius.

The data discussed above have all related to the center of the rotating spoke. Figure 17 is a display of ion density as a function of azimuth,

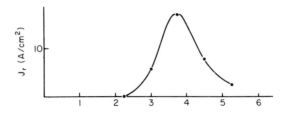

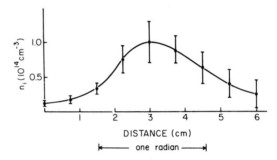

Fig. 17. Azimuthal distribution of inferred ion density at $r = 3$ cm. The current spoke is included for reference.

derived from the time dependence of the forward probe signal at $r = 3.0$ cm, and including the assumption that v_θ is the same at all azimuths. It will be recalled that the **E** × **B** measurements support this assumption.

The figure contains the noteworthy result that the current spoke and the plasma spoke are not exactly identical, although their peaks roughly coincide. Substantial plasma moves ahead of the current. (Note the arrow indicating the direction of spoke motion.)

16.3.3.3.3. DISTRIBUTION OF v_r. Since the azimuthal flux observations support a simple Langmuir-type interpretation of ion current data, one may turn the differential flux probe toward other directions in order to obtain a full specification of the plasma velocity vector.

In this series of experiments, v_r was examined next. Since the probe was left at the same position, but merely rotated 90°, the assumption could safely be made that the same $Zn_i e$ applies to measurements along r as well as along θ. Thus, the velocities are simply in the same ratio as the probe currents. Figure 18 is a display of v_θ, v_r, and v_z (also obtained in the same fashion). The current spoke is not drawn here, but its peak is at 3.7 cm on the azimuthal distance scale, and indicated by a tic on the bottom abscissa.

A remarkable feature of the v_r distribution is that the plasma counter-

16.3. ELECTRIC PROPULSION AS A PLASMA PHYSICS PROBLEM

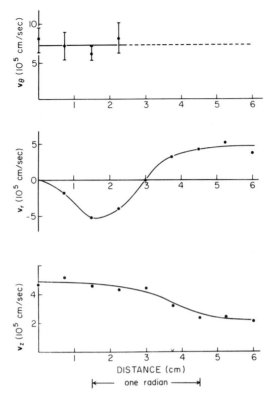

FIG. 18. Inferred azimuthal radial and the axial plasma velocities as a function of azimuth; $r = 3$ cm, $z = 10$ cm.

streams radially, with the separation between opposite radial flows placed near the leading edge of the current spoke. Ahead of the current, the plasma streams outward, as would be expected in the absence of radial forces. Within the current channel however, ions move radially inward, and thus participate in the current flow. It is found by this means that ions carry about 30% of the total current at the radius of this particular measurement. The anode sheath, mentioned earlier (Section 16.3.3.2.2) is adequate to reflect inward those ions which stream outward in advance of the spoke. There must also exist within the anode sheath an azimuthal plasma current such that $j_\theta B_z$ is sufficient to transfer momentum radially at the required rate. This layer of current has been observed.[13]

16.3.3.4. Resistivity of Plasma. It was mentioned in Section 16.3.3.2.3 that the triangular points which lie in the spoke region of Fig. 14 correspond to a particular assumed value of η, the plasma resistivity; the fit was performed by the use of Eq. (16.3.7), assuming that $\mathbf{v}_\theta$ is that of the current spoke. Similarly, Eq. (16.3.8) was fitted to the data, and a value of η in close agreement with the previous one was obtained.

When this resistivity was compared with that computed from Spitzer's formula, using $T_e = 1.1$ eV measured spectroscopically, it was found that

$$\eta_{\text{obs}} = 4.0(\pm 1.5)\eta_{\text{sp}},$$

where η_{sp} is the Spitzer resistivity. It seems likely that collective plasma phenomena are responsible for raising resistive losses above their classical thermal values.

16.3.3.5. *Azimuthal Electric Field.* It will be noted from Eq. (16.3.9) that E_θ can be derived from other measured quantities, since this component does not involve the plasma resistivity η. Hence, a comparison between this value and that measured with a differential floating probe affords an excellent check on internal consistency of the diagnostics. Figure 19 shows

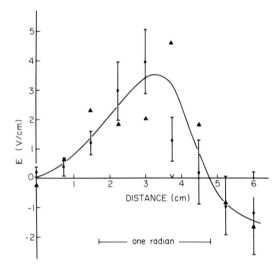

FIG. 19. The azimuthal distribution of $\mathbf{E}_\theta$. The points and error bars represent values measured with a differential probe, and triangles are values derived from Ohm's law.

both distributions, the points (with scatter bars) representing direct measurements, while the triangles are derived values. The only azimuth at which the two determinations seem to differ significantly is at the position of the spoke center (as indicated by the tic mark). In general, the agreement is excellent, considering the rather circuitous derivation through Eq. (16.3.9).

16.3.3.6. Summary of Data. The MPD arc experiment as described in the foregoing sections involves the intercombination of many separate pieces of data as obtained by different diagnostic methods. In order to summarize the whole process, the flow charts of Fig. 20 have been constructed.

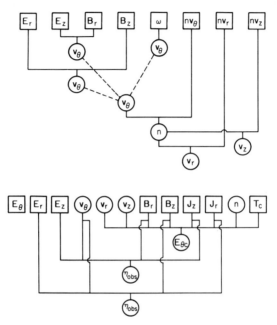

FIG. 20. Flow charts illustrating the inference of plasma parameters from measurements. Squares represent direct measurements, while circles enclose inferred quantities.

In these charts, quantities in square boxes are directly measured; quantities in circles are inferred. The top chart shows first that the azimuthal plasma velocity is established by the agreement of three independent measurements. The flux probe then allows a solution for the ion density, which in turn makes possible the evaluation of v_r and v_z, from flux probe data taken along those directions.

In the lower chart, we see how the plasma resistivity and E_θ are derived. Here, some of the input data are themselves inferred quantities. Yet the fact that $E_{\theta c}$ (calculated) agrees quite closely with E_θ (measured) validates these inferences.

AUTHOR INDEX

Numbers in parentheses are reference numbers and indicate that an author's work is referred to although his name is not cited in the text.

A

Agafonov, V. I., *187*, *188* (11), *190* (11), *215* (11), *224* (11)
Alcock, A. J., *24*
Allen, C. W., *12*
Alpher, R. A., *3*, *11*, *12*, *13* (1)
Angerami, J. J., *94*, *100*, *101*, *102* (13), *106*, *107*, *108*, *111*, *115*, *116* (22)
Artsimovich, L. A., *204*
Asbridge, J. R., *85*, *88*
Ascoli-Bartoli, U., *3*, *13* (6), *23*, *26*, *27* (22, 23)
Ashby, D. E. T. F., *33*, *260*, *261*
Astrom, E., *136*

B

Baconnet, J. P., *234*
Bader, M., *55*
Baker, D. A., *33*, *35* (44), *211*
Balsley, B. B., *179*, *183*, *185*
Bame, S. J., *49*, *50*, *52* (2), *55*, *77* (7), *85*, *88*
Banks, P. M., *107*
Bannerman, D. E., *215*
Barber, P. M., *27*
Barkhausen, H., *89*
Baron, M. J., *172*, *184*
Barrington, R. E., *113*, *117*, *120*, *131*, *134*
Barron, D. W., *161*, *162* (28), *172* (28)
Beams, J. W., *3*
Beckner, E. L., *187*, *227* (8)
Belrose, J. S., *113*, *117*, *120*
Bennett, S., *270*
Bennett, W. H., *210*
Bernfeld, M., *156*, *157*
Bershader, D., *3*
Beynon, J. D. E., *39*

Bezrukikh, V. V., *52*, *55*
Bhatnagar, P. L., *163*
Binsack, J. H., *55*, *58*, *69*, *70* (13), *71*
Birks, L. S., *225*
Black, N. A., *265*
Bogen, P., *30*
Bohm, D., *282*
Bonetti, A., *55*, *60*
Booker, H. G., *178*
Born, M., *20*
Bostick, W. H., *196*
Bottoms, P. J., *187*, *188*, *189*, *193*, *198* (7), *205* (7), *224* (19), *230*, *238* (23), *247* (7)
Bowles, K. L., *142*, *155*, *159*, *178* (3), *179*, *180*, *181*, *182*
Brace, L. H., *175*
Brice, N. M., *106*, *107* (17), *108* (17), *109*, *117*, *118*, *119* (26), *120*, *122*, *123*, *126* (32), *128* (33), *129*, *130*, *131* (35), *136*
Bridge, H. S., *53*, *55*, *59*, *60* (10), *83*, *84* (15)
Brooks, R. E., *42*, *47*
Budden, K. G., *141*
Buges, J. C., *47*
Buneman, O., *165*, *184*
Burlaga, L. F., *85*
Butler, T. D., *44*, *215*, *248*

C

Cain, J. C., *178*
Cantarano, S., *72*
Carlson, H. C., *168*
Carpenter, D. L., *89*, *94*, *100*, *101*, *103*, *104*, *105*, *106*, *107*, *108*, *109*, *110*, *113*, *115*
Carpenter, J. P., *188*, *189*, *193*, *224* (19), *230*, *238* (23)
Carru, H., *168*
Casale, P., *196*

Cesari, G., *234*
Chamberlain, J., *39*
Cipolla, F., *187*, *188* (12), *190* (12), *222* (12), *224* (12)
Clothiaux, E. J., *227*
Cohen, M. H., *161*
Cohen, R., *179*, *180*, *181*, *182*
Comisar, G. G., *187*, *188* (10), *222* (10), *227* (10), *243*
Conte, S. D., *163*
Cook, C. E., *156*, *157*
Coudeville, A., *234*
Craig, J. P., *30*, *33*
Critchfield, C. L., *215*

D

Dacus, E. N., *268*
Damon, K. R., *51*
Dangor, A. E., *39*
Davenport, W. B., Jr., *146*
Davis, J. M., *55*, *59*, *83*, *84* (15)
Davis, L. R., *69*
DeAngelis, A., *13*
Decker, G., *187*
Dellis, A. N., *30*
DeMichelis, C., *24*
Deshmukh, A., *196*
Dilworth, C., *53*, *60* (10)
Dougal, A. A., *30*, *33*, *46*, *187*
Dougherty, J. P., *161*, *162* (28), *163*, *164*, *171*, *172* (28, 36), *185*
Ducati, A. C., *269*
Dunckel, N., *113*
Dungey, J. W., *99*
D'iachenko, V. F., *187*, *188* (11), *190* (11), *215* (11), *224* (11), *247*
Dyson, P. L., *175*

E

Earl, W. H. F., *30*
Eckdahl, C., *271*, *281* (13), *287* (13)
Eckersley, T. L., *95*
Egidi, A., *55*, *72*
Emmert, G., *196*
Enemark, D. C., *64*, *67*
Evans, D. S., *72*
Evans, J. V., *147*, *150*, *160*, *168*

F

Falconer, I. S., *30*
Farber, E., *196*
Farley, D. T., *148*, *149*, *150*, *151* (10), *152*, *155* (10), *160*, *161*, *162* (28), *164*, *165* (29), *166*, *170*, *171*, *172* (28, 35), *175* (51), *184*, *185*
Fejer, J. A., *161*
Filippov, N. V., *187*, *188* (11), *189*, *190*, *203*(5), *215* (11), *224* (11), *247* (5), *248*
Filippova, T. I., *187*, *188* (11), *189* (4), *190* (11), *203* (5), *215* (11), *224* (11), *247* (5), *248* (4)
Finkelstein, D., *27*, *33* (34)
Formisano, V., *84*
Fowler, T. K., *243*
Frank, L. A., *55*, *64*, *67*, *69*
Freeman, J. W., Jr., *55*, *72*, *73*
Fremouw, E. J., *168*
Fried, B. D., *163*
Friedman, A. M., *188*
Freidrich, O. M., Jr., *46*, *187*
Fünfer, E., *15*, *18* (15)

G

Gabel, R. H., *64*, *67*
Garriott, O. K., *177*
Garwin, R., *200*, *203* (33), *210* (33)
Gates, D. C., *187*, *188* (14), *196*, *224* (14)
Gebbie, H. A., *39*
George, A., *39*
Gerardo, J. B., *33*, *37*
Giannini, G. M., *269*
Gibson, A., *37*
Gilbert, H. E., *85*, *88*
Glasstone, S., *10*, *214*, *229* (40)
Golub, G. V., *187*, *188* (11), *190* (11), *215* (11), *224* (11)
Golubchikov, L. G., *187*, *188* (11), *190* (11), *215* (11), *224* (11)
Gooding, T. J., *260*, *261*
Gordon, W. E., *158*
Gourlan, C., *187*, *188* (12), *190* (12), *222* (12), *224* (12)
Grassmann, P. H., *30*
Gribble, R. F., *30*, *31*, *33*
Griem, H. R., *228*
Gringauz, K. I., *49*, *50*, *52*, *55*
Gross, E. P., *163*

AUTHOR INDEX

Grover, F. W., *268*
Grunberger, L., *196*
Gurnett, D. A., *121*, *123*, *124* (34), *125*, *126* (32), *128*, *129*, *130*, *131* (35)

H

Haegi, M., *187*, *188* (12), *190* (12), *222* (12), *224* (12)
Hagfors, T., *161*, *173*
Hain, K., *15*, *18* (15)
Haines, K. A., *46*
Hall, L. A., *51*
Hamal, K., *24*
Hamilton, W. R., *96*
Hammel, J. E., *33*, *35* (44)
Hanson, W. B., *175*
Hasselman, K., *155*
Hayworth, B. R., *260*, *261*
Heflinger, L. O., *42*
Helliwell, R. A., *96*, *101*, *104*, *113*
Henderson, N. K., *64*, *67*
Henins, I., *44*, *215*, *248* (41)
Herman, J. R., *176*
Herold, H., *15*, *18* (15), *33*, *36* (48)
Hildebrand, B. P., *46*
Hinrichs, C. K., *187*, *188* (14), *196*, *224* (14)
Hinteregger, H. E., *51*
Holder, D. W., *26*
Holzer, T. E., *107*
Hora, H., *234*
Hundhausen, A. J., *49*, *52*, *55*, *70* (1), *77* (1, 7), *82*, *85*, *88*
Hunt, A. G., *18*

I

Igenbergs, P., *15*, *18* (15)
Imshennik, V. S., *187*, *188* (11), *190* (11), *215* (11), *224* (11), *247*
Irving, J., *27*
Itoh, Y., *47*
Ivanov, V. D., *187*, *188* (11), *190* (11), *215* (11), *224* (11)

J

Jacobson, L., *55*
Jahn, R. G., *251*, *255*, *265*
Jahoda, F. C., *26*, *33*, *35* (44), *36* (48), *41*, *43* (60), *44*, *46*, *48*, *215*, *248* (41)

Jeffries, R. A., *41*, *43* (60), *45*
Jelly, D., *109*
Jephcott, D. F., *33*
Jermakian, A., *196*
Johnson, F. S., *139*
Johnson, W. B., *37*
Johnson, W. C., *120*
Jones, D., *126*
Jones, I. R., *268*
Josias, C., *71*

K

Kakos, A., *47*
Keepin, G. R., *249*
Keller, J. B., *164*
Khrabrov, V. A., *187*
Kiefer, J. H., *24*
Kinder, W., *13*, *15* (14), *17* (14)
King, P. G. R., *31*
Kittredge, R. I., *53*, *68* (12)
Knöös, S., *27*
Knox, F. B., *184*
Knudsen, W. C., *53*
Kolesnikov, U. A., *187*, *188* (11), *190* (11), *215* (11), *224* (11)
Komisarova, I. I., *45*
Kon, S., *39*
Korobkin, V. V., *38*
Kribel, R., *271*, *281*, *287* (13)
Krook, M., *163*
Küpper, F. P., *15*, *18* (15)
Kunze, H. J., *3*

L

Laaspere, T., *120*
Ladenburg, R., *3*
Lafferty, D. L., *187*, *188* (14), *196*, *224* (14)
Lanter, R. J., *215*
Larsen, A. B., *37*
Larson, A. V., *260*, *261*
Lawrence, J. L., Jr., *71*
Lazarus, A. J., *55*, *59*, *60*, *83*, *84* (15), *85*
Leadabrand, R. L., *184*
LeBlanc, J., *244*, *247* (61)
Lehner, G., *215*, *219*
Leith, E. N., *39*
Lewis, R. L., *164*
Liebing, L. F. C., *269*

Linhart, J. G., *187*, *188* (12), *190* (12), *222* (12), *224* (12)
Little, E. M., *18*, *26*, *30*, *31*
Long, J. W., *187*, *228*, *229*, *230*, *231*, *232* (9)
Longley, J., *207*
Longmire, C. L., *188*, *207*, *243* (17), *244*
Lovberg, R. H., *10*, *23*, *24* (24), *25*, *27* (21), *214*, *229* (40), *258*, *271*, *281* (13), *287* (13)
Lutz, R. W., *24*
Lyon, E. F., *53*, *55*, *60* (10), *71*, *83*

M

McClure, J. P., *175*
MacCullagh, J., *96*
MacDonald, G., *155*
McEwen, D. J., *131*, *134*
McIlwraith, N., *53*, *55*, *68* (11), *72* (11)
McKay, K. G., *52*
McKenzie, A. S. V., *27*
McKibbin, D. D., *55*, *64*, *84* (24)
Maguire, J. J., *73*
Maissonnier, C., *187*, *188* (12), *190*, *222*, *224* (12)
Malein, A., *30*
Malyutin, A. A., *38*
Marconero, R., *72*
Marshall, J., *44*, *189*, *215*, *248* (41)
Martellucci, S., *3*, *10* (5), *11*, *12* (5), *13* (5), *23*, *26*, *27* (22, 23)
Mason, R. H., *55*, *64*, *84* (24)
Mather, J. W., *187*, *188* (6), *189*, *192*, *193* (27), *194*, *198* (2, 7, 27), *203* (6), *205* (7), *215* (2), *222* (6), *224* (19), *230* (2), *234* (6), *238* (23), *244* (2), *247* (7, 54)
Matsushita, S., *179*
Mayhall, D. J., *187*
Mazzucato, E., *23*, *26*
Medford, R. D., *18*
Medley, S. S., *33*
Medved, D. B., *61*
Meehan, R. J., *27*
Melnikov, V. V., *55*
Meskan, D. A., *187*, *188* (10), *222*, *227* (10)
Middleton, D., *154*
Montgomery, M. D., *52*, *55*, *77* (7)
Moorcroft, D. R., *165*, *170*, *171*
Morgan, M. G., *120*
Morozov, A. I., *204*, *249*
Morse, R. L., *30*, *31*, *44*, *215*, *248* (41)

Muehlberger, E., *269*
Munk, W., *155*
Murphy, B. L., *52*
Muzzio, J., *134*
Myers, M. A., *55*

N

Neher, L. K., *215*
Nelms, G. L., *120*
Ness, N. F., *85*
Neugebauer, M., *55*, *67*, *84*
Nishida, A., *106*
Nishijima, Y., *27*
North, R. J., *26*

O

Ochs, G. R., *179*
Ogilive, K. W., *53*, *55*, *68* (11, 12), *72* (11), *85*
Olbert, S., *58*, *70* (14), *77* (14), *85* (14), *86*
Osher, J. E., *189*
Oster, G., *27*
Ostrovskaya, G. V., *3*, *45*, *47*
Ostrovsky, Yu. I., *3*, *47*
Otsuka, M., *39*
Ozerov, V. D., *52*, *55*

P

Paolini, F. R., *65*, *81*
Park, C. G., *101*, *103*
Parker, L. W., *52*
Parkinson, G. J., *39*
Patou, G., *187*, *205* (15), *223*, *224* (15)
Peacock, N. J., *187*, *188*, *224* (20), *228* (9), *229* (9), *230* (9), *231* (9), *232* (9)
Perkins, F. W., *167*, *168*, *169*
Petit, M., *168*
Petriceks, J., *168*, *172*
Petrov, P. P., *187*
Pizzella, G., *72*
Poehler, T. O., *39*
Poeverlein, H., *96*
Pohl, F., *215*, *219*
Potter, D., *247*
Poros, D. J., *178*
Powell, A. L. T., *18*
Presby, H. M., *27*, *33* (34)
Prior, W., *196*

Q

Quinn, W. E., *18, 26, 30, 31*

R

Ramsden, S. A., *30*
Randall, R. F., *64, 67*
Ratcliffe, J. A., *149*
Razier, M. D., *189*
Reid, G. W., *37*
Renau, J., *161*
Ribe, F. E., *26*
Richardson, H., *85*
Rishbeth, H., *177*
Roberts, K. V., *242, 247*
Robouch, A., *187, 188* (12), *190* (12), *222* (12), *224* (12)
Root, W. L., *146*
Rosenbluth, M. N., *161, 173, 174, 200, 203* (33), *210* (33)
Rossi, B., *53, 55, 60* (10)
Rostoker, N., *161, 173, 174*
Rusbüldt, D., *30*
Rybchinskii, R. E., *52, 55*
Rye, B. J., *27*

S

Salpeter, E. E., *161, 167* (24)
Samuelli, M., *187, 188* (12), *190* (12), *222* (12), *224* (12)
Savenko, I. A., *55*
Savin, B. I., *55*
Sawyer, G. A., *18, 26, 41, 43* (60)
Scarf, F. L., *84*
Scherb, F., *53, 55, 60* (10), *84*
Schlobohm, J. C., *184*
Seasholtz, R. G., *161, 165* (21)
Shapiro, L. L., *45*
Shawhan, S. D., *123, 124* (34), *125, 126* (32), *128*
Sherman, A., *256*
Siemon, R. E., *18*
Silva, R. W., *55, 64, 84* (24)
Simonett, A., *187, 205* (15), *223* (15), *224* (15)
Singer, S., 55, *72*
Siscoe, G. L., *85*
Skadron, G., *186*
Smith, D. R., *227*
Smith, H. M., *39*
Smith, R. L., *89, 96, 97, 99* (10), *111, 113, 115, 116* (22), *117, 118, 119* (26), *123, 126* (32), *136*
Snyder, C. W., *55, 67, 83, 84*
Sosnowski, T. P., *37*
Sperli, F., *72*
Spitzer, L. Jr., *5, 204, 243* (36), *244* (36), *278*
Stanley, W. W., *64, 67*
Stern, M. O., *268*
Steward, G. J., *31*
Stone, K., 108, *109*
Storey, L. R. O., *89*
Stratton, T. F., *229*
Strausser, Y. E., *61*
Strong, I. B., *85, 88*
Stuhlinger, E., *251*
Sugiura, M., *178*
Sutton, F. W., *256*
Suydam, B. R., *242, 244*
Svirsky, E. B., *187, 188* (11), *190* (11), *215* (11), *224* (11)
Swift, D. A., *27*

T

Tanenbaum, B. S., *161, 165* (21), *171, 278*
Taylor, J. B., *242*
Theodoridis, G. C., *62, 81*
Thomas, K. S., *36*
Thompson, W. B., *278*
Thomson, J. J., *158*
Tozer, D. A., *27*
Tsuruta, T., *47*
Tsytovich, V. N., *189, 247* (21)
Tuck, J. L., *223*
Turner, R., *39*

U

Untiedt, J., *178*
Unwin, R. S., *184*
Upatnicks, J., *39*

V

Van Allen, J. A., *69*
Van Paassen, H. L., *187, 188* (10), *222* (10), *227* (10)

Van Vleck, J. H., *154*
Vasyliums, V. M., *55*, *61*, *84* (17), *85* (17)
Verdeyen, J. T., *33*, *37*
Vernov, S. N., *55*
Villars, F., *142*
Vinogradov, V. P., *187*, *189* (4), *218* (5)

W

Waldteufel, P., *184*
Walkup, J. F., *113*
Waller, J. W., *27*
Ward, S., *30*
Ware, K. D., *188*, *189*, *193*, *224* (19), *230*, *238* (23)
Warren, C. S., *73*
Watteau, J. P., *187*, *205* (15), *223* (15), *224* (14), *234*
Weigh, F., *46*
Weinreb, S., *154*
Weinstock, J., *186*
Weisskopf, V. F., *142*
Weyl, F. J., *3*
White, D. R., *3*, *11*, *12*
Wilcox, P. D., *187*, *228* (9), *229* (9), *230* (9), *231* (9), *232* (9)
Wilkerson, T. D., *53*, *55*, *68* (11, 12), *72* (11)
Williams, A. H., *188*, *189*, *192*, *193* (27), *194*, *198* (27), *224* (19), *230*, *238* (23)
Williamson, J. H., *33*
Williamson, J. M., *69*
Wilson, R., *244*, *247* (61)
Wolf, E., *20*
Wolfe, J. H., *55*, *64*, *84* (24)
Wolter, H., *21*, *27*
Woodman, R. F., *164*, *172* (37), *173*
Wright, J. K., *18*
Wuercker, R. F., *42*
Wulff, H., *30*

Y

Yabroff, I., *96*
Yamanaka, M., *39*
Yngvesson, K. C., *168*
Yoshinaga, H., *39*

Z

Zaidel, A. N., *3*, *45*, *47*
Zorskie, J., *196*

SUBJECT INDEX

A

Afternoon "bulge," whistlers and, 110
Antennas, for whistlers, 90–91

B

Bistatic measurement of spectrum and autocorrelation function of radio waves, 150–153
Bow shock, definition of, 50

C

Cauchy formula, 12
Ciné-interferometry, plasma applications of, 46–47
Collision frequency, 110
Continuous channel multiplier, in deep space plasma measurements, 72
Coupled cavity interferometry, 31–35
  application to pulsed plasmas, 32–33
  frequency response, 33–35
  sensitivity of interferometer for, 32
Coupling phenomena, of whistlers, 124–126
Cowling conductivity, 177
Curved-plate analyzer(s)
  balanced type, 63
  comparison of instruments, 68–70
  coupled energy-angular response, 65–67
  crossed-field velocity selector, 68
  for deep space plasma studies, 62–70
  energy-angle asymmetry ("skewing"), 65
  extraneous effects, 68
  geometries of, 63–64
  types of, 64
  unbalanced type, 63
Cylindrical curved-plate analyzers, 64

D

Debye length, of a plasma, 49
Deep space plasma measurements, 49–88
  experiments involving, 1959–1968, 54–55
  of flux, 71–73
    AC methods, 71–72
    DC methods, 71
    counting methods, 72–73
    synchronous detector, 71–72
  hydrodynamic parameters, 77
  instrumentation for, 52–76
    calibration of, 736
    curved-plate analyzer, 62–70
    modulated-potential Faraday cup, 53–62
  methods of analysis, 77–88
    differential and integral behavior, 80–82
    distribution function by interpolation, 85–88
    equations for detector current, 79
    estimation of plasma parameters, 83
    measurement relation to particle distribution function, 78–80
    method of moments, 83
    model distributions, 83–85
  microscopic character of instruments, 49–50
  organization of energy and angle measurements, 73–76
    on spin-stabilized spacecraft, 74–75
    on triaxially stabilized spacecraft, 75–76
  plasma properties, 49
  plasma regions, 50
  spacecraft-plasma interaction problems and, 51–52

Dense plasma focus, 187–249
  apparatus for, 190–194
    coaxial plasma focus, 190
    low-inductance cables, 192
    vacuum spark gap, 193
      conditioning
  computer study, 244–248
  development of, 194–215
    acceleration phase, 198–203
    breakdown phase, 194–198
    collapse phase, 203–204
    filamentary structures, 196
    sheath resistance, 204
    "snowplow" model, 200
  dynamic behavior of current sheath, 205–208
  energy partition in discharge, 209–210
  impedance considerations, 210–215
  plasma diagnostic measurements, 215–249
    accelerated deuteron beam model, 219
    deuterium-tritium experiment, 222–224
    electron temperature, 229–233
    neutron production, 215–222
    neutron scaling, 224–225
    spatial stability of focus, 238–242
    X-ray emission, 225–229
  theoretical study, 242–244
Dispersion relation, 6, 9–11
Double pulse method for measurement of spectrum and autocorrelation function of radiowaves, 150
Ducting, of whistlers, 95–98

E

Echo trains, of whistlers, 92–93
  duct of, 9
  electron cyclotron frequency, 93
  earth-ionosphere wave-guide, 93
Eckersley law, in ducting of whistlers, 95
Eikonal equation, 20
Electrical propulsion
  criteria for, 251–252
  as a plasma physics problem, 269–289
    azimuthal electric field, 288
    electric field measurement, 277
    ion flux measurements, 282
    MPD arc parameters, 270
    potential contour map, 280
    preheated cathode, 271
    Rogowski coils, 272–276
  plasma problems in, 257–289
  as a problem in magnetohydrodynamics, 255–269
  systems for,
    categories of, 254–255
    plasma parameters in, 252–254
Electrojet, equatorial, scattering from, 176–186
Electromagnetic engines, 255
Electron cyclotron frequency, 9
Electron plasma frequency, 9
Electrothermal engines, 254

F

Faraday rotation, in optical refractivity of plasmas, 27–31
  magnitude of, 27–29
  measurement of, 29–31
Flux, deep space measurements of, 71–73
Flux density, in deep space plasma measurements, 80
Four-pulse method for, measurement of spectrum and autocorrelation function of radio wave scattering, 152

G

Gas laser interferometry, 31–39
  coupled cavity interferometry, 31–35
  off-axis modes, 37
  oscillating mirror, 36–37
  rotating feedback mirror for, 35–36
  without coupling of laser and plasma, 38–39
Gordeyev integral, 164
Group velocity, 11

H

Hall mobility, 177
Helium whistlers, 131–133
Hemispherical curved-plate analyzer, 64
Heterodyne interferometry, 37–38
  principle of operation, 37
  sensitivity, 37–38

SUBJECT INDEX

Holographic interferometry, plasma
  applications of, 42–48
  diffuser in scene beam, 42–43
  introduction of background fringes,
    43–44
  two-color type, 45–46
Hydrodynamic parameters, of space
  plasmas, 77

I

Ion-cutoff whistlers, 134–135
Ion engines, 254–255
Ion flux measurements, 282–287
  Langmuir probe application, 282–283
Ionospheric radio wave scattering,
    139–186
  from a diffuse medium, 142–158
    theory, 142–146
  electron density fluctuations, 142
    autocorrelation function, 144
    Fourier transform of, 143
  from equatorial electrojet, 176–186
    irregularities, 183
    VHF scatter observations from,
      178–179
  frequency stepping, 156
  Gaussian random variable, 145
  higher-order spectral functions, 155–156
    bispectrum, 155
  incoherent type, 158–176
    plasma line, 167–169
    power spectrum, 166–167
    theory, 160–164
    total scattering cross section, 165–166
  measurement techniques, 146–158
    for polarization, 149
    for power, 147–148
    for spectrum and autocorrelation
      function, 150–153
  parameters of, 140
  power spectrum, 144
  pulse compression, 156–158
    ambiguity function, 157
  pulse repetition frequency (PRF), 156
  refractive index fluctuations, 144
  scattered power, 143
  true height profiles, 141
  virtual height profiles, 141

K

Knee whistlers, 103–107
  in plasmapause, 102–103

L

Langmuir probe applications, in ion flux
  measurements, 282–283

M

Mach-Zehnder interferometers, 13–20
Magnetohydrodynamics (MHD),
    electromagnetic propulsion as a
    problem in, 255–269
  distributed currents, 266
  electrical efficiency, 259
  energy inventory in a coaxial gun, 260
  linear channel, 257–260
  magnetic probes, 258–260
  particle analyzers, 268
  pulse-forming network, 261
  transducers, 268
Magnetopause
  definition of, 50
  plasma parameters of, 51
Magnetospherically-reflected whistlers,
    115–117
Magnetosheath, definition of, 50
  plasma parameters of, 51
Magnetotail, definition of, 50
Modulated-potential Faraday cup
  angular response, 58–59
  in deep-space plasma measurements,
    53–62
  diagram of, 56
  energy window of, 57
  extraneous effects, 60–62
  high-order corrections for, 57–58
  transmission function, 58
  "Venetian blind" collimeter of, 59
Multiple-ion plasmas, wave propagation
  in, 135–137

N

Nose whistlers
  frequency and time delay of, 93
  propagation delay of, 93
Nyquist theorem, 162

O

Optical interferometry, 13–20
  Mach-Zehnder interferometer for, 13–20
    effect of bending of the rays, 18–20
    fringe localization, 15
    light sources, 15–18
  Schlieren methods, 20–22
    derivation of angular deflection, 21–22
    detection of angular deflection, 22–27
    double-inclined-slit type, 24–25
    grid projection technique, 27
    phase contrast methods, 27
    sensitivity limits, 27
    shadowgrams, 25–27
Optical refractivity of plasmas, 1–48
  ciné-interferometry, 46–48
  experimental methods requiring laser sources, 31–48
  gas laser interferometry, 31–39
  Faraday rotation, 27–31
  heterodyne interferometry, 37–38
  holographic phase measurements, 39–48
  noninterferometric methods for, 47–48
  standard experimental methods, 13–31
    optical interferometry, 13–20
  theory of, 3–13
    nonelectronic refractivity contributions, 11–13
    wave equation, 3–9

P

Pedersen mobility, 177
Plasma
  in deep space, *see* Deep Space plasma measurements
  optical refractivity of, 1–48
Plasma detectors, for spacecraft, calibration, 73
Plasma frequency, definition of, 6
Plasma refractive index, definition of, 6
Plasma sheet
  definition of, 50
  plasma parameters of, 51
Plasmapause, knee whistlers in, 102–103
Plasmaphere, definition of, 50
Polarizability, 12
Postacceleration detection of deep space plasma, 72
Proton whistler, 120–131

Q

Quadrispherical curved-plate analyzer, 64

R

Radio wave scattering, in ionosphere, 139–186
Rogowski coils, 272–276
Rotation matrix, in deep space plasma measurements, 79

S

Satellites, whistler observation by, 111–137
  ducted whistlers, 111–112
  ion effects, 112–120
Schlieren methods, optical interferometry, 20–22
Shadowgrams, in Schlieren methods, 25–27
Snell's law, in whistler ducting, 95
Solar wind, plasma parameters of, 51
Spacecraft, plasma interaction problems of, 51–52
  photoelectrons from spacecraft, 51
  secondary electrons, 51
Subprotonospheric whistler, ion-effect observations on, 112–115

T

Transmission function, in deep space measurements, 79

V

"Venetian blind" collimeter, of modulated-potential Faraday cup, 59

W

Wave equation, 3–9
  solution for collisions, 6–7
  solution for no collisions, 5–6

SUBJECT INDEX

solution with static applied magnetic field, 7–9
Whistlers, 89–137
  afternoon "bulge" and, 110
  antennas for, 90–91
  broadband installation for, 91
  ducting of, 95–98
    Eckersley law, 95
    lifetimes of, 95
    satellite observation of, 111–112
    Snell's law, 96
    total internal reflections and, 97
  dynamic spectra of, 91
  echo trains of, 92–93
    causative atmosphere, 92
  electric field measurements of, 108–110
  electron density distribution along field lines, 107–108
  equatorial electron density of, 98–103
    diffusive equilibrium model of electron density, 99
    gyrofrequency model of electron density, 99
    R^{-4} model and, 100
  experimental methods for, 90–91
  flux-tube content of, 98–103
  fractional hop type, 124–126
  ground-based observations on, 92–110
  helium type, 131–133
  "impulse response" and, 90
  ion-cutoff type, 134–135
  knee type, 103–107
  lightning discharges and, 89
  lower hybrid resonance noise of, 117–120
    in nonducted mode, 117
  magnetospherically-reflected type, 115–117
  mode waves of, refractive index, 96
  multiple ion effects of, 120–124
  nose type, 93–95
  "one-hop" type, 93
  propagation delay of, 89
  proton type, 120–124
    magnetic field strength and, 127–128
    proton number density of, 128–129
    proton temperature and, 129–131
  signal of, minimum detectable power of, 91
  subprotonospheric type, ion effects, 112–115
  "two-hop" type, 93